CONTENTS

Tectonomagnetics and Local Geomagnetic Field Variations

ADVANCES IN EARTH AND PLANETARY SCIENCES

Advances in Earth and Planetary Sciences 5

Special Issue of Journal of Geomagnetism and Geoelectricity

Tectonomagnetics and Local Geomagnetic Field Variations

Proceedings of IAGA/IAMAP Joint Assembly
August 1977, Seattle, Washington

Edited by
M. Fuller
M. J. S. Johnston
T. Yukutake

Springer Science+Business Media, B.V.

ORIGINALLY PUBLISHED BY CENTER FO ACADEMIC PUBLICATIONS JAPAN IN 1979
SOFTCOVER REPRINT OF THE HARDCOVER 1ST EDITION 1979

ISBN 978-94-010-9827-4 ISBN 978-94-010-9825-0 (eBook)
DOI 10.1007/978-94-010-9825-0

Preface

Physical and chemical studies of the earth and planets along with their surroundings are now developing very rapidly. As these studies are of essentially international character, many international conferences, symposia, seminars and workshops are held every year. To publish proceedings of these meetings is of course important for tracing development of various disciplines of earth and planetary sciences though publishing is fast getting to be an expensive business.

It is my pleasure to learn that the Center for Academic Publications Japan and the Japan Scientific Societies Press have agreed to undertake the publication of a series "Advances in Earth and Planetary Sciences" which should certainly become an important medium for conveying achievements of various meetings to the academic as well as non-academic scientific communities. It is planned to publish the series mostly on the basis of proceedings that appear in the Journal of Geomagnetism and Geoelectricity edited by the Society of Terrestrial Magnetism and Electricity of Japan, the Journal of Physics of the Earth by the Seismological Society of Japan and the Volcanological Society of Japan, and the Geochemical Journal by the Geochemical Society of Japan, although occasional volumes of the series will include independent proceedings.

Selection of meetings, of which the proceedings will be included in the series, will be made by the Editorial Committee for which I have the honour to work as the General Editor. I and the members of the Editorial Committee will certainly welcome any suggestions that will promote the series. Whenever the convener of a meeting related to earth and planetary sciences is in a position to have to look for a medium for publishing the proceedings please contact us.

Tsuneji Rikitake
General Editor

Foreword

The fields of tectonomagnetism and tectonoelectricity have assumed an important role in the complex subjects of fault mechanics and earthquake prediction. Unfortunately, not all aspects of these measurements are well understood. A symposium on this subject was held at the I.A.G.A. Third General assembly on August 23, 1977, in Seattle, U.S.A. Eleven of the eighteen papers presented at this symposium appear in completed form in this special issue of *Journal of Geomagnetism and Geoelectricity*. Two papers were rejected. The abstracts of all papers can be found in *EOS*, **58**, 730–733, 1977.

The papers naturally fall into three main groups: (1) Observation, interpretation and limitations in the measurements of local and regional magnetic fields; (2) Magnetotelluric experiments and results; (3) Laboratory experiments of the effects of stress on the magnetic properties of rocks.

The editors of this proceedings issue would like to thank the reviewers: R.F. Butler, P.E. Davis, R. Day, T.L. Henyey, R.L. Kovach, E.R. Niblett, R.L. Parker, J. Revol, P.N. Shive, B.E. Smith, and F.D. Stacey for their prompt response and Diane Mondragon for preparing the final manuscripts of several of the papers.

Michael Fuller
Malcolm Johnston
Takeshi Yukutake

Guest Editors

Symposium on Tectonomagnetics and Small Scale Secular Variations Held at the IAGA/IAMAP Joint Assembly at Seattle on Tuesday, August 22nd, 1977

Co-convenors: V.A. Shapiro* and M.J.S. Johnston**

**Institute of Geophysics of U.S.S.R., Sverdlovsk, U.S.S.R.*
***U.S. Geological Survey, Menlo Park, California, U.S.A.*

The symposium was held under the auspices of Division 1 of the IAGA and consisted of a morning and afternoon session. The various papers covered geomagnetic field observations, magnetotellurics, resistivity studies and also laboratory analyses of the effect of uniaxial and hydrostatic stress upon magnetization.

Amongst the papers describing geomagnetic field precursors, Johnston's paper describing the model of stress accumulation in the region of the locked part of the San Andreas in the general region of San Juan Bautista is of particular interest and makes testable predictions. The paper by Shapiro and Abdullabekov describing the failure to observe a magnetic field precursor, or indeed any effect at the time of the Gazly earthquake, was discouraging in view of the magnitude of the event and the proximity of the magnetometers to the epicenter. In contrast, the results reported from Middle Asia included more positive indications of precursors. In the U.S.A., Smith and others report correlation of changes in creep rate with magnetic events. One hopes that an unambiguous precursor will be seen shortly. Otherwise serious questions concerning the potential value of the method will have to be faced.

The magnetotelluric results also seem to be encountering the same difficulty in defining unambiguous precursors, although Miyakoshi and Suzuki appear to have detected an anomalously high conductivity region possibly related to the Yoshioka-Shikano fault in Japan east of Matsue. The important paper by Kurtz and Niblett demonstrates that the electrical properties are indeed strongly time-dependent with changes of up to 30% in the impedence tensor taking place in a few months. In addition, a steady change has also taken place. Again it remains to be seen whether an association between seismic activity and the changes can be established.

The laboratory experiments by Henyey and others and by Revol and others clearly establish that stress brings about important changes in both induced and remanent magnetization. Much of the interest at present is to develop techniques which can give the temperatures, confining pressures and uniaxial stress to provide a realistic simulation of the effects going on at depth in the crust. In the unconfined experiments some remarkable changes have been seen, which are both anomalous and also potentially useful as precursors.

This symposium once again made clear that the electromagnetic methods have a very long way to go before they can be regarded as reliable earthquake precursors with practical value. Nevertheless there are indications that with proper instrumentation the effect may be within the range of possible observations. The development of three component magnetometer networks with low drift rates is particularly important not only in the interpretation of any precursors which may be seen, but also in enhancing signal to noise discrimination.

Tectonomagnetic Studies in Tajikistan

Yu. P. Skovorodkin, L.S. Bezuglaya, and T.V. Guseva

Institute of Physics of the Earth, Academy of Sciences U.S.S.R.,
Moscow, U.S.S.R.

(Received October 10, 1977)

High-precision simultaneous observations of the geomagnetic field have been made over areas with fault displacements and junctions of mountain structures in Tajikistan using proton magnetometers MPP-1 with 0.1 γ sensitivity. The optimal duration and frequency of resurveying as well as the detection level of possible tectonomagnetic variations (0.6 γ) have been experimentally determined.

Examples of characteristic local geomagnetic variations possibly associated with changes of tectonic stresses are given. Oscillations with periods of one or a few weeks and amplitudes of 0.7–1.5 γ occur during seismically quiet times, and monotonic variations continuing over a month with subsequent change of sign and a tendency for rapid recovery to the former level occur shortly before energy class $K>11$ earthquakes (on the Rautian scale, where $K=1.8M_L+4$).

1. Introduction

In recent years it has been demonstrated that the geomagnetic field measurements can be used to study stresses in rocks of the upper crust. This is borne out by magnetic observations during industrial explosions, near reservoirs, and around epicentral areas in the vicinity of active faults (Abdullabekov *et al.*, 1976; Kozlov *et al.*, 1974; Johnston *et al.*, 1976; Skovorodkin *et al.*, 1971). This affords an opportunity to study stress changes which may be related to tectonic activity.

In a number of cases premonitory changes in the local magnetic field have been detected at large epicentral distances (Skovorodkin *et al.*, 1971). The amplitude of these changes is largely dependent on focal depth and earthquake magnitude, as well as on the relative positions of observation sites and the directions of the geomagnetic field and stress axes. Thus, for a given region a small earthquake having an epicenter favorably situated with regard to the observation sites may produce a sharply-defined change in the magnetic field while a larger one having a differently situated hypocenter may not.

An important in these investigations is the use of high-sensitivity absolute magnetometers, thus making possible the detection of tectonomagnetic variations with amplitudes of one to several gammas. However, at the same time, these instruments greatly enhanced the 'noise' contribution from inductive currents in conductively inhomogeneous crust. One must have a long series of systematic observations in a particular area, using a variety of geophysical methods, to be able to decipher tectonomagnetic effects. Examples of such studies are observations in the western United States on the active San Andreas fault zone, and investigations in the Tajikistan area at the junction of the Pamirs and Tien-Shan mountain structures (the Garm prognostic testing ground).

2. Tectonomagnetic Observations

Repeated magnetic field observations have been made since 1974 at a fixed-station network with instruments both in mountains and river valleys. The sites were situated in a variety of tectonic environments. Measurements were taken simultaneously in the field at permanent sites, being synchronized within 0.3 sec. Distances between the permanent site and the resurveyed network sites ranged between 1 and 30 km.

We used MPP-1 type proton magnetometers with sensitivity of 0.1 γ and absolute error in the measurement of T of $\pm 0.2\,\gamma$.

Making simultaneous observations at two or more sites as suggested by A.G. Kalashnikov (1954) excludes geomagnetic variations of extraterrestrial origin; however there still remains a part of internal origin, due to induction currents. Inhomogeneities in electrical conductivity at shallow depths may give rise to local irregularities in the geomagnetic variation at neighboring sites. We know (see, for example, Johnston *et al.*, 1976; Rikitake, 1966) that geomagnetic variability due to induction currents, ΔT_i, may amount to several gammas between sites up to 10 km apart. Obviously, even for small site separations, an appreciable portion of ΔT_i is contributed by short-period geomagnetic variation. This can be reduced by averaging the measured values of T over a time interval comparable with the characteristic periods of oscillation. Thus, for short-period variations of the Pc-type a 10-min interval can be considered sufficient. In addition, for the averaging to be satisfactory and the random error to be reduced to a minimum, one must determine the optimum number of independent observations, n.

First we chose a value $\Delta\chi$ such that at a certain confidence level it will be less than the systematic error δ. One puts as a rule $\Delta\chi<\delta/3$ or $\Delta\chi<\delta/2$. Then provided $\Delta\chi=\delta/2$ and $s_n=\delta$, where s_n is the mean square error of an individual observation, we need at least $n=18$ observations to make the error less than 1.5δ at a confidence level $\alpha>0.95$.

The systematic error is determined by the instrumental error. For fixed-array simultaneous observations, that is, magnetometers functioning as variometers with 0.1 γ sensitivity, the maximum error in ΔT for a single observation is $\delta=\pm 0.2\,\gamma$. When one magnetometer is in fixed-array operation and the other records at a network of sites in repeated surveys, $\delta=\pm 0.3\,\gamma$.

Since it is impossible to treat in a rigorously quantitative manner disturbing factors such as non-stationarity of variations with periods of greater than 600 sec ('bays,' S_q—and other variations), determination of 'noise' level for the purpose of detecting possible effects associated with piezomagnetic properties of rocks must be carried out experimentally in each particular investigation area. Assuming that in seismically quiet time there is no tectonomagnetic variation present during a 24-hr period, the actual mean square error of measurement unconnected with the piezomagnetic effect may be estimated by a method similar to that in use in magnetic prospecting (LOGACHEV, 1951). It is

$$M=\sqrt{\frac{\sum^{m}(\delta\Delta T)^2}{2m}}$$

where $(\delta\Delta T)_m=\Delta T_m-\Delta T_1$ is the difference in amplitude between the first and m-th measurement cycle, and m is the number of repeated cycles. Mean 10-min differences have been taken as ΔT_m; time intervals between measurement cycles ranged from several minutes to several hours. The value of M was ± 0.1 and $\pm 0.2\,\gamma$, for distances between sites (L) equal to 200 m and 5–10 km respectively.

We see that the experimental error does not exceed the error of the method in both cases. Hence we conclude that the 10-min averaging procedure is satisfactory, and the 'noise' component due to electrical conductivity inhomogeneities in the surrouding rocks does not contribute significantly to the variation of ΔT.

The mean square error of a single observation of ΔT for a 10-min interval is $s_n \leqq \pm 0.2\,\gamma$ when $20<n<30$.

Thus for the area under investigation, tectonomagnetic variation may be judged significant when the change in ΔT exceeds $0.6\,\gamma$.

3. Observations

During the observation time concerned, the following earthquakes have occurred in the Garm testing grounds: four earthquakes of $K=13$ energy class, a sequence of $K=11$, and the large Isfarin-Batken earthquake ($K=15.8$) whose epicenter lies 135 km from the permanent site. There have also been long periods of seismic quiescence (for months at a time).

We now describe a few typical cases of geomagnetic field variation at sites in the area investigated.

1) The Simiganch profile lies along the Hissar deep-fault zone. There is geological evidence for a fault lying close to Sites 3 and 4 (Fig. 1), along a contact between Paleozoic granites on the one hand and Mesozoic sandstones, aleurolites, and limestones on the other. The maximum difference of local field values along the profile line from the permanent site (0) to Site 7 was found to be $85\,\gamma$.

Figure 1 shows also the variation of the difference $\Delta T=T_0-T_i$ ('i' is the site number) for a period of over 3 weeks. We see that field level oscillations have been observed at all sites (with respect to the permanent one), with a period of about 2 weeks. The oscillation amplitude at Site 7 reached $1.5\,\gamma$.

In addition, records at Sites 3 and 4 near the fault zone reveal a steady shift in the mean level of ΔT. However, no significant seismic events occurred within a distance of several tens of kilometers during the observation time nor during the

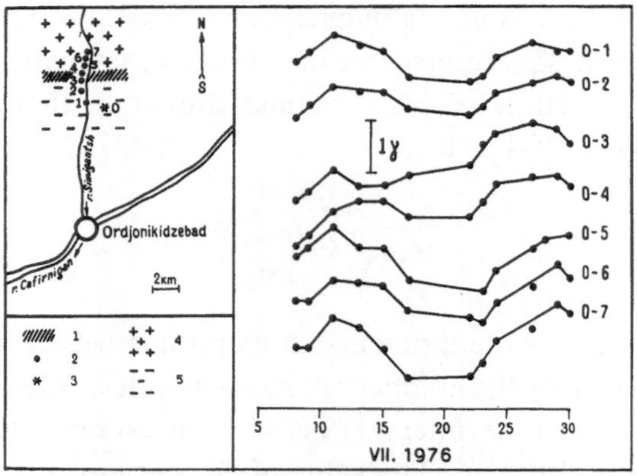

Fig. 1. Variation of ΔT and a scheme of the Simiganch profile. 1, fault zone; 2, sites; 3, the permanent site; 4, granite (Pz); 5, sandstone, etc. (Mz).

previous and following few months. The observed oscillations of ΔT may be associated with changes in tectonic stresses over the area, which must be especially pronounced in the proximity of a fault zone.

2) Figure 2 shows the positions of the fixed sites set up over various tectonic units within a portion of the junction zone between the Pamirs mountain system and the southern Tien-Shan, along the Surhob fault line. Three of the sites (1, 3179, 3040) as well as the permanent site 41, are all situated within the Mandapul block composed of Paleozoic granite. Site 2 is situated at the southern end of the Kabut-Krym mountains, Site 2589 is in Cretaceous sediments in the Pamirs front zone.

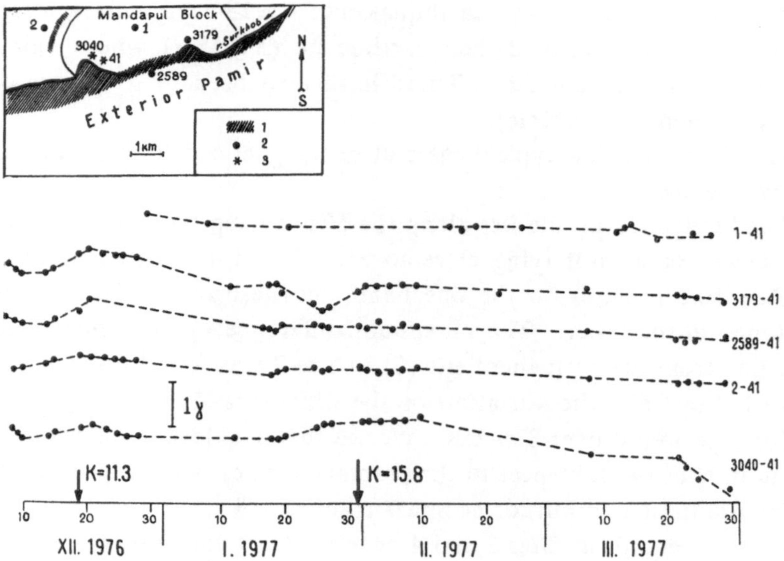

Fig. 2. Variation of ΔT and a scheme of the Garm profile. 1, fault zone; 2, sites; 3, permanent sites.

This figure also shows the variation of $\Delta T = T_1 - T_{41}$ for a period of 3.5 months. During this time a local earthquake ($K=11.3$) occurred (focal depth $H=9$ km, epicentral distance $L=6$ km, azimuth $A=103°$), as well as the Isfarin-Batken earthquake ($H=25$ km, $L=135$ km, $A=15°$). The origin times of the two earthquakes are shown by arrows in the figure.

We thus see that the variations of ΔT are different even within the single Mandapul block. The maximum variation was observed over the eastern portion of the block (Site 3179), and the minimum at the center (Site 1). Site 2589 south of the fault zone did not record any variation of ΔT. Superposed upon the general background in the variation of ΔT there were relatively short period magnetic field changes taking place a short time before the earthquakes and probably associated with pre-earthquake stress redistribution processes. A characteristic feature of these changes was a fairly long monotonic decrease of the field (Site 3179, from the end of December, 1976 through January 26, 1977) combined with subsequent rapid increase (during 6 days). The variation of ΔT in December, 1976 may have been associated with the local earthquake of December 18.

We thus see that there are individual features in the variation at sites that are close to one another but lie within different parts of tectonic structures.

3) The Obi-Hingous profile cuts across the main structures of the Peter I

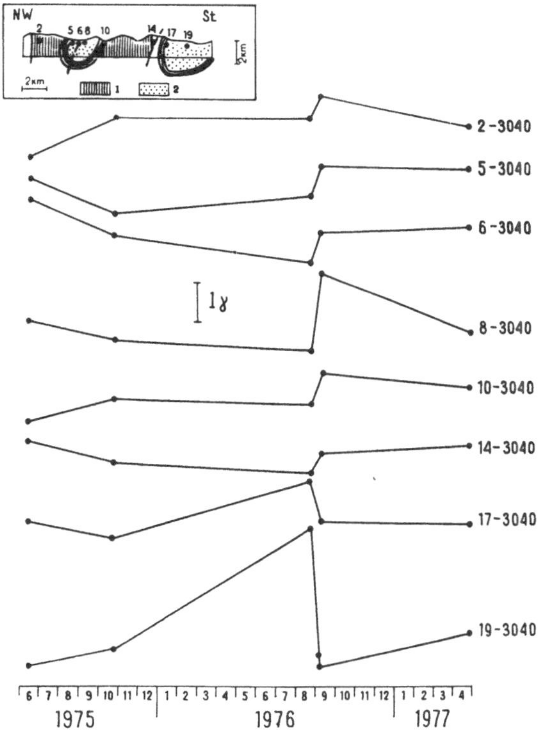

Fig. 3. Variation of ΔT and a schematic cross-section of the Obi-Hingous profile. 1, clay deposits; 2, sandstone.

Range consisting of sand and clay deposits (Fig. 3). A series of measurements was made on August 26, 1976, shortly before a $K = 13.2$ earthquake (September 3) and another on September 6–9.

We see that the greatest variations (up to $3\,\gamma$) have been observed at Sites 8 and 19, over the central portions of the synclinal folds, while the changes at the other sites have been rather small, although all the sites are at the same epicentral distance.

4. Conclusions

High-precision resurvey magnetic measurements made in seismically active areas in Tajikistan have revealed local geomagnetic variations with an amplitude of up to $3\,\gamma$, associated in all probability with changes of tectonic stresses in rocks. The variation may be different in type:

1) Oscillations with a period of a week or more unconnected with the occurrence times of earthquakes.

2) Monotonic changes of level during a considerable time with subsequent change of sign shortly before earthquakes (for $K > 11$).

3) Monotonic changes of the field (for months at a time).

The geomagnetic variations are significantly different both in amplitude and direction at different adjoining or closely situated geologic structures, even for teleseismic events. This is probably due, not only to the magnetic, but also to the rheological properties of rocks.

The authors gratefully acknowledge their obligations to academician M.A. Sadovsky, Director, Institute of Physics of the Earth, I.L. Nersesov, Head of Seismological Sector, Institute of Physics of the Earth, and to S. Kh. Negmatullaev, Director of the Tajik Institute of Seismology and Antiseismic Construction, for their valuable advice and help with the work.

REFERENCES

Abdullabekov, K., Kh. Abdullaev, E. Berdaliev, V. Shapiro, and A. Pushkov, Local variations of the geomagnetic field over the Charvak reservoir area, Main Geomagnetic Field and Problems of Paleomagnetism, summaries of reports, Part I, 3, 1976.

Johnston, M.J.S., B.E. Smith, and R. Mueller, Tectonomagnetic experiments and observations in Western U.S.A., *J. Geomag. Geoelectr.*, **28**, 85–97, 1976.

Kalashnikov, A.G., Potentialities of magnetometric methods for the problem of earthquake forerunners, *Trudy Geofiz. Instit.*, No. 25 (152), 162–180, 1954.

Kozlov, A.N., A.N. Pushkov, R. Sh. Rakhmatullin, and Yu. P. Skovorodkin, Magnetic effects from explosions in rocks, *Izv. AN USSR, Ser. Physics of the Earth*, No. 3, 66–71, 1974.

Logachev, A.A., *A Course of Magnetic Prospecting*, pp. 303, Gosgeolizdat, Moscow, 1951.

Rikitake, T., Elimination of non-local changes from total intensity values of geomagnetic field, *Bull. Earthq. Res. Inst.*, **44**, 1041–1070, 1966.

Skovorodkin, Yu. P., L.S. Bezuglaya, and V.N. Vadkovsky, Magnetic measurements over epicentral zones, in *Experimental Seismology*, edited by M.A. Sadovsky, pp. 398–402, Nauka, Moscow, 1971.

An Attempt to Observe a Seismomagnetic Effect during the Gazly 17th May 1976 Earthquake

V.A. SHAPIRO* and K.N. ABDULLABEKOV**

*Institute of Geophysics, U.S.S.R. Academy of Sciences, Sverdlovsk, U.S.S.R.
**Institute of Seismology, Uzbek Academy of Sciences, Tashkent, U.S.S.R.

(Received September 1, 1977)

Portable proton magnetometers with a sensitivity of 0.1 nT operated before, during and after the moment of the magnitude $M=7.3$, 17th May 1976, Gazly earthquake. Magnetometers were operating essentially at the epicenter at the Kara-Kyr station and 170 km to the northeast at the Tamdybulak. Measurements were made during the period 13–22 May 1976. No variations of the geomagnetic field total force with an amplitude exceeding 0.1 nT were seen during the earthquake. Similar results were obtained from the magnetic measurements during 12 aftershocks with $M \geqq 4.5$ which followed the main quake. The profiles of the changes of the geomagnetic field during these 12 aftershocks were stacked. The stacked plot shows no variations with amplitude greater than 0.1 nT.

1. Introduction

Magnetic variations were observed with portable proton magnetometers during the period from 13 to 22 May 1976 at the epicenter of the 17th May 1976 Gazly earthquake. The magnetometer T-MП (MAKSIMOVSKIKH and SHAPIRO, 1976) operated throughout this time at the Kara-Kyr point which was the epicenter of the earthquake. From the 13th, until the moment of the main shock on the 17th of May, measurements of the total force modulus of the geomagnetic field were made 3 times every 10 min. The sensitivity was 0.06 nT and the averaging time 30 sec. After the time of the magnitude $M=7.3$ earthquake (02 hr 58 min 38 sec universal time) these measurements were taken every minute at the epicenter. The readings were taken uninterruptedly during periods of aftershocks. From the time of the main shock on the 17th of May continuous observations of T, at the maximum rate of 6 readings a minute, were made for 30 min. We also managed to record the geomagnetic field during several aftershocks, with an experimental model of the T-MП magnetometer, with a reading accuracy of 0.015 nT and a measuring time of 4 sec.

During this same period of time another proton magnetometer was operating at Tamdybulak, 170 km to the northeast of the epicenter. The observations on this instrument were taken every 10 min, synchronized with the magnetometer at Kara-Kyr to within 1–2 sec. The third proton magnetometer operated during the main

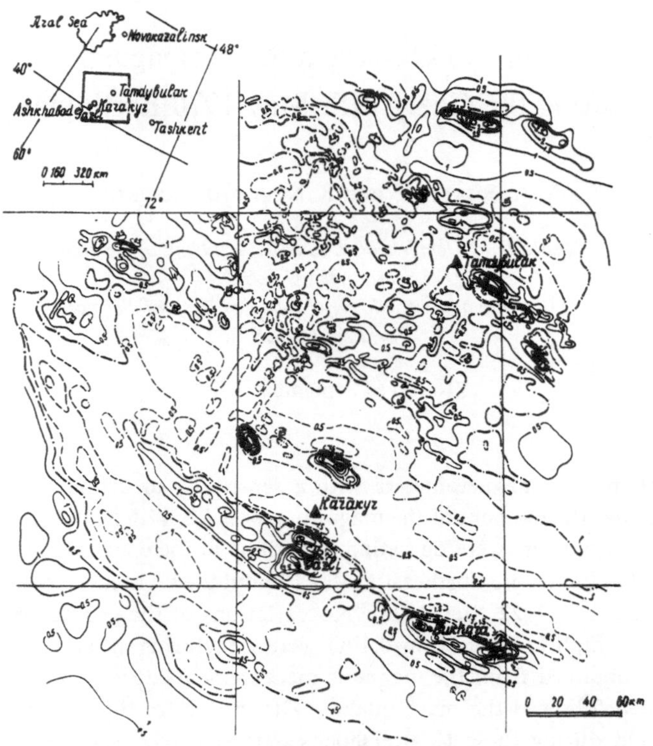

Fig. 1. Magnetometer locations and aeromagnetic map of the total force T, showing the epicenter (▲) of the Gazly 1976 earthquake.

shock and 20 min after it at point No. 58 of the Kyzyl-Kum polygon, 20 km to the west of Tamdybulak.

Measurements at all 85 points of Kyzyl-Kum polygon began in 1974 and 1975 and were repeated just after the earthquake. The polygon, represented with 5 profiles, is situated at the distance of 130–220 km to the north and northeast of the earthquake epicenter. To help identify possible magnetic field effects of the earthquake the data from stationary magnetic observatories Yangi-Bazar (Tashkent), Vannovskaya Ashkhabad and Novokazalinsk were utilized. Thus, the geomagnetic field data used for the present work consists of the data of measurements of T with the proton magnetometers at three points during the earthquake, variational investigations at the observatories and repeated observations at the polygon (Fig. 1).

2. Experimental Data

Processes generating local variations of the geomagnetic field, the sources of which are placed in the upper parts of the lithosphere, are divided into three types according to their temporal characteristics (GOLOVKOV *et al.*, 1977).

1) Most rapid—the time of their existence never exceeds a few days;
2) Rapid—with the duration of their existence from a week to a few years;

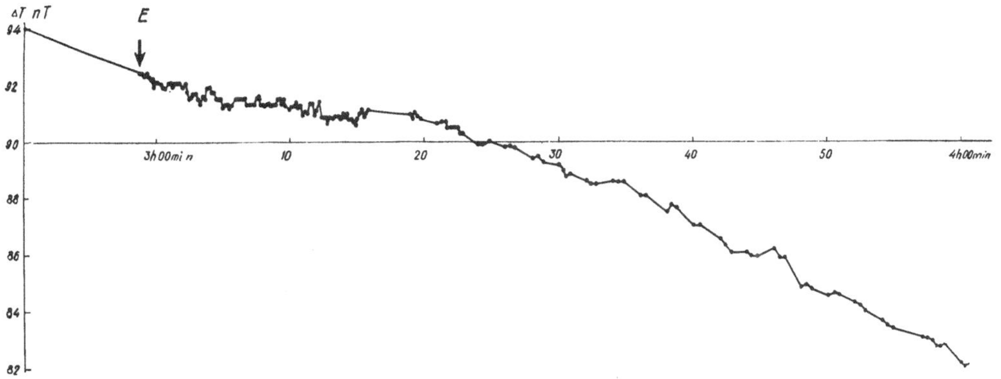

Fig. 2. Total force T variations recorded at the epicenter of the earthquake (point Kara-Kyr). E: moment of the shock with magnitude M=7.3.

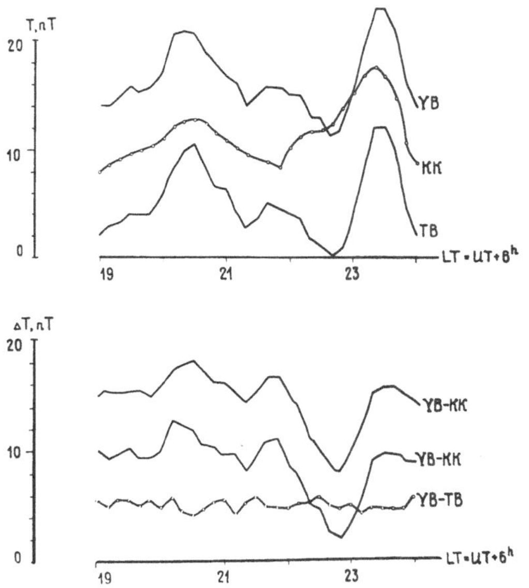

Fig. 3. Total force T curves, 15th May 1976. YB, Yangi-Bazar (Tashkent); TB, Tamdybulak; KK, Kara-Kyr.

3) Slow—with the duration of a dozen years and even more.

Measurements with the proton magnetometers can be made at a maximum rate of one reading per 10 sec (6 sec—to polarize a sample, 2 sec—to count and 2 sec—to record a reading). Thus, this type of measurement gives information which has a cut off within the most rapid processes such that if the duration is less than 30 sec it will not be well determined.

Plotting of the field T recorded at the point Kara-Kyr (Fig. 2) shows that no variations of the field T with a duration exceeding 30 sec and amplitude more than 0.1 nT, which could be interpreted as a seismomagnetic effect, were registered on the 17th of May at the moment of the main shock. The same negative conclusion can be made about the magnetic field recorded during the 12 strongest aftershocks with magnitude $M \geqq 4.5$.

Comparison of variations in T recorded at Kara-Kyr, Tamdybulak and observatories Yangi-Bazar, Vannovskaya and Novokazalinsk showed that variations taking place during the magnetic field information investigations at Kyzyl-Kum were identical at Tamdybulak, Yangi-Bazar, Vannovskaya and Novokazalinsk. The changes of the field at Kara-Kyr were practically similar. The only exception is the 15th May 1976 bay-like disturbance of the field which occurred identically at all the points of observations with the exception of Kara-Kyr, where this variation took an anomalous form (Fig. 3).

3. Discussion

We should like to interpret the 15th May disturbance as a phenomenon linked to the earthquake, because it occurred within 29 hr of the earthquake. The phenomenon of anomalous passage of a bay-like disturbance was known long ago and is relatively well studied (ROKITYANSKY, 1975). Analyzing the anomalous passage of this type of variation it is possible to determine the presence of conductivity zones in the earth's crust. The zones of anomalous conductivity in the U.S.S.R. are being studied both in seismic areas and in essentially aseismic ones. Thus, the anomaly recorded at Kara-Kyr during the bay-like disturbance of the field may be evidence for a conductivity anomaly near the epicenter of the earthquake. It supports the presence of currently active geological structures at the epicenter. On the other hand, we cannot entirely exclude the possibility that there is not only information about active structure but also information about forthcoming earthquakes as well in the anomaly of the 15th of May bay. Unfortunately, at this time we can only state that these observations support the data of geologists and geophysicists about the presence of active fractures in the Kara-Kyr area. Further research may help to explain in detail the 15th of May anomaly and to distinguish information on a presage, if it exists. To accomplish this, two kinds of research should be done: (1) Repeating every 2–3 weeks the observations of T and other components of the geomagnetic field at Kara-Kyr and Tamdybulak; (2) holding of the profile investigation at the epicenter area according to the standard method (GOLOVKOV *et al.*, 1977), which will enable us to contour the conductive zone and to study, together with variational investigations in the anomaly of the region, the character and peculiarities of the anomaly.

The variations in magnetic field T recorded during the time of 12 aftershocks with $M \geq 4.5$ are summed in order to emphasize any seismomagnetic signals. This is done with the assumption that any aftershock brings useful information of itself in the geomagnetic field, but the amplitude of that useful signal is weak and is lost in the background of the geomagnetic noise. The following procedure was used: The aftershock time and the field amplitude at the moment of the shock were assumed equal to zero. We have considered the variations for the interval of ± 5 min around the aftershock. The results of this treatment of the data are presented in Fig. 4. Evidently, at the moment of shock there is no visible effect with an am-

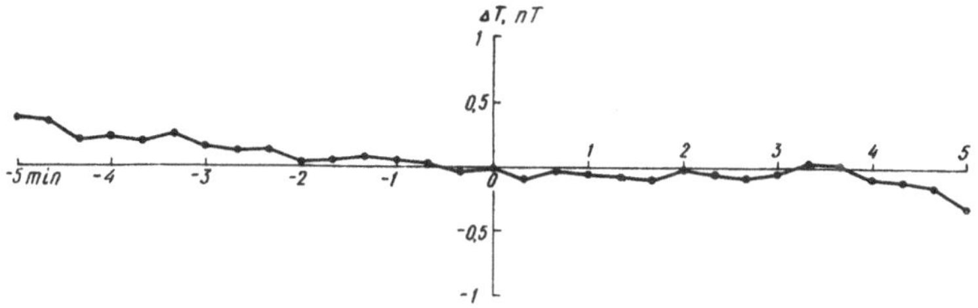

Fig. 4. Summary total force T curve for 12 aftershocks of the Gazly 17th May 1976 earthquake.

plitude exceeding 0.1 nT. The repeated observations of T at the Kyzyl-Kum polygon showed that there was not any observed anomaly exceeding the observation error, which might be considered as a phenomenon connected with the earthquake.

4. Conclusions

Thus, summarizing the results of the investigations and their initial interpretation we can come to the following conclusions:

1) No magnetic effect with an amplitude exceeding 0.1 nT was observed either for the main shock ($M=7.3$) or for the aftershocks during the period 17th–22nd May 1976. This conclusion does not refer to possible variations with periods less than 30 sec which cannot be detected by the instruments utilized. Such effects are observed in the laboratories (IVANOV *et al.*, 1973) and in the field during explosions (KOZLOV *et al.*, 1974).

2) We have not observed any steady drift of the field.

3) We did observe an anomalous passage of a bay-like disturbance of the field 29 hr before the magnitude 7.3 earthquake of the 17th of May which may indicate the presence of a conductive zone in the epicenter area. The dependence of the degree of this disturbance anomaly and the forthcoming earthquake is not established. Further research is underway to see whether the described anomalous bay contains information of the forthcoming earthquake.

4) In accordance with the analysis of the observational data at the Kyzyl-Kum polygon no visible anomalous changes in the field T were revealed either before or after the earthquake.

We thank N.V. Fedorovskaya, E.B. Berdaliev, M.U. Muminov, F.D. Narmirzaev for their dedicated help while the field observations took place and S.I. Maksimovskikh as well for the development of the equipment for the experiment.

REFERENCES

GOLOVKOV, V.P., N.A. IVANOV, I.M. PUDOVKIN, and V.A. SHAPIRO, *Instruction on the Searches and Investigations of Secular Variation Anomalies of the Geomagnetic Field*, 4–24, Nauka, Moscow, 1977.

IVANOV, N.A., V.A. SHAPIRO, and A.A. NULMAN, Investigation of a rate of rock samples dynamic magnetization in constant geomagnetic field, Paleomagnetism and Rock Magnetism Symposium, Naukova Dumka, Kiev, 152–153, 1973.

KOZLOV, A.N., A.N. PUSHKOV, R.Sh. RAKHMATULLIN, and U.P. SKOVORODKIN, Magnetic effects while the explosions in the rocks, *Izv. AS USSR, Fisika Zemly*, No. 3, 66–71, 1974.

MAKSIMOVSKIKH, S.I. and V.A. SHAPIRO, Portable proton magnetometer of high sensitivity, *Geomagnetism y Aeronomia*, **16**, 2, 389–391, 1976.

ROKITYANSKY, I.I., *Investigation of Electroconductivity Anomalies by the Method of Magnetovariational Profiling*, 32–46, Naukova Dumka, Kiev, 1975.

Secular Variation Anomalies and Aseismic Geodynamic in the Urals

V.A. SHAPIRO, A.L. ALEINIKOV, A.A. NULMAN,
V.A. PYANKOV, and A.V. ZUBKOV

Institute of Geophysics, Ural Scientific Center, U.S.S.R. Academy of Sciences, Sverdlovsk, U.S.S.R.

(Received September 1, 1977)

Secular variation anomalies (SVA) of two types were revealed as a result of high-precision measurements of the geomagnetic field in the Urals. The results suggest significantly non-uniform tectonic stresses in the Urals. An attempt to measure the time dependence of stress variations was made. An anomaly of the first type is characterized by the steady growth of the field at a rate of 2–3 nT/year. Laboratory experiments with the rocks showed that similar changes might be caused by the changes of magnetic susceptibility and remanent magnetization with changes of hydrostatic pressure of about 10 bars. The anomalies of second type border the zone of folded Urals to the East and the West. They exhibit sign changes in trends of up to ± 20 nT/year. These anomalies are probably caused by the change of the rock conductivity because of changes in pore pressure in the zones of thrusting.

1. Introduction

Within the Urals region, high-precision measurements of the total geomagnetic field T have been made on a grid with points situated in different tectonic zones. Synchronous observations were made with a number of stationary and temporary magnetic observatories as the points of reference (Fig. 1). This permits the reduction of systematic errors of observation up to 0.2–1.2 nT (SHAPIRO and IVANOV, 1974). The aim of the research is to determine secular variation anomalies (SVA) and to study their relation to geodynamic processes, taking place in the earth's crust and upper mantle. Such processes in tectonically active zones of the lithosphere may give rise to a local distortion of the normal pattern of geomagnetic field variations observed on the earth's surface (SHAPIRO, 1976).

Several mechanisms for the generation of anomalous geomagnetic variations are possible. Rock magnetic studies have shown that thermal and chemical effects, constant and alternating magnetic fields, electric currents and ionizing radiation, static and dynamical burdens all change the magnetic state of a rock and consequently give an anomalous magnetic field around it (NAGATA, 1961; TRUKHIN, 1973). As piezomagnetic, thermomagnetic and the various other effects are fundamentally different in origin, the amplitudes, periods and characteristic sizes of the anomalies, created by them, may also be different.

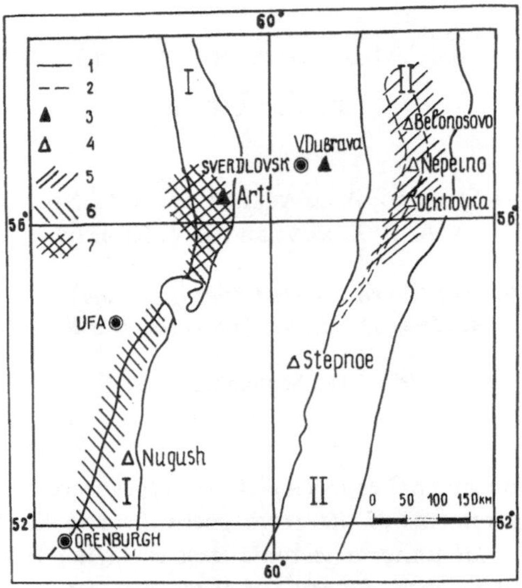

Fig. 1. Secular variation anomalies (SVA) at the Urals. 1, borders of geological structures of the first order (I, Preduralsky trough; II, Zaurals uplift); 2, borders of geological structures of the second order (Talitza megasynclinorium); 3, permanent magnetic observatories; 4, temporary magnetic observatories; 5, Butka SVA; 6, Bashkir SVA; 7, Manchazh SVA.

Investigations of possible physical mechanisms which could give rise to anomalous geomagnetic variations at the earth's surface, have been carried out. First, we consider the magnetic state of the rock and its response to changes in pressure, temperature, chemical environment (STACEY *et al.*, 1965; SHAPIRO, 1968; BARSUKOV *et al.*, 1968). Some anomalies may be explained by these effects. However, to explain the wide spectrum of anomalies observed, an additional approach to the events under consideration appears to be required. A potentially important mechanism is the movement of fluid in networks of capillaries in rock, accompanied by conductivity changes. The surface magnetic effects are commensurate with piezo- and thermomagnetic ones, under otherwise identical conditions (SOBOLEV *et al.*, 1975; MIZUTANI and ISHIDO, 1976).

2. The Results of the SVA Investigation

In order to study contemporary processes in the lithosphere, a highly accurate geomagnetic survey of the total force (T) of the geomagnetic field has been made in the Urals since 1968. The survey covers territory of 400,000 km^2 within the limits of the Southern and Middle Urals. Thirty-five hundred long-term observation sites were set up. In this region anomalous time-dependent field changes were discovered. The geomagnetic field T secular variation anomalies, revealed at the Urals (SHAPIRO and IVANOV, 1974; SHAPIRO and PYANKOV, 1976) may be divided by their spatial and temporal characteristics into two types.

2.1 *Anomalies of the first type*

The Manchazh regional anomaly, which is characterized by a monotonic growth of the field T up to 2–3 nT/year, is of the first type. This anomaly was discovered in 1969 between the Arti observatory and the Manchazh settlement, that is 140 km to the West of Sverdlovsk. This anomaly is situated at the Eastern margin of the Russian platform which forms a protrusion, bounded by the Urals structure, within the Middle Urals. The secular variation anomaly coincides with a stationary anomaly of the magnetic field T of intensity 1,200 nT. Detailed investigations of the nature of the changes of the normal field showed that the sign of the time derivative of the anomalous field intensity does not depend on the sign of the same derivative of the normal field (Fig. 2). This result suggests that the nature of Manchazh regional anomaly is in part linked to changes in magnetization of rock at depth and not simply due to the induced effect of the secular variation (BULASHEVITCH and SHAPIRO, 1975).

2.2 *Anomalies of the second type*

The Bashkir and Butka secular variation anomalies (Fig. 1) are characterized by sign changes in trends of the total force T of the geomagnetic field with amplitudes up to 20 nT/year. The background of normal changes is ± 2 nT/year. These are considered to be anomalies of the second type. The Bashkir secular variation anomaly is situated in the Preduralsky trough and coincides with the Western border of the folded Urals. This anomaly is fixed by seven latitudinal profiles to the South from the Ufa city to the latitude of the Orenburgh city and by several profiles to the North from the Ufa city. The anomaly is seen in two submeridional zones with a width of about 20 km each and a length of 500 km within the limits of the studied territory (Fig. 3).

In the zone of Zilair megasynclinorium, which is 80–100 km to the East from the Bashkir SVA, in 1975–1976 an elongated SVA was discovered. It is analogous

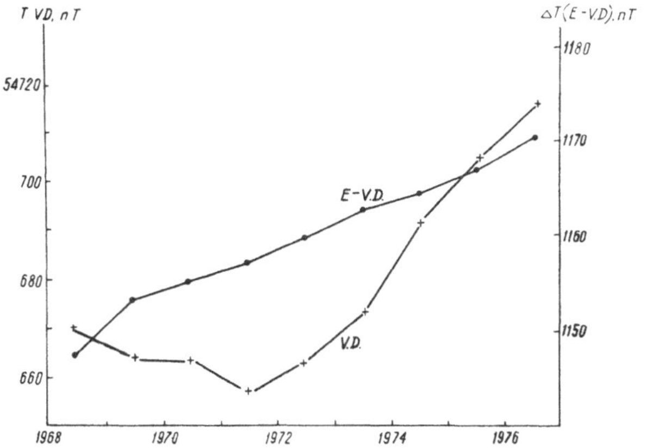

Fig. 2. Secular variation of geomagnetic field total force T in observatory V. Dubrava (curve with crosses) and difference curve epicenter of Manchazh anomaly $-$V. Dubiava (curve with circles).

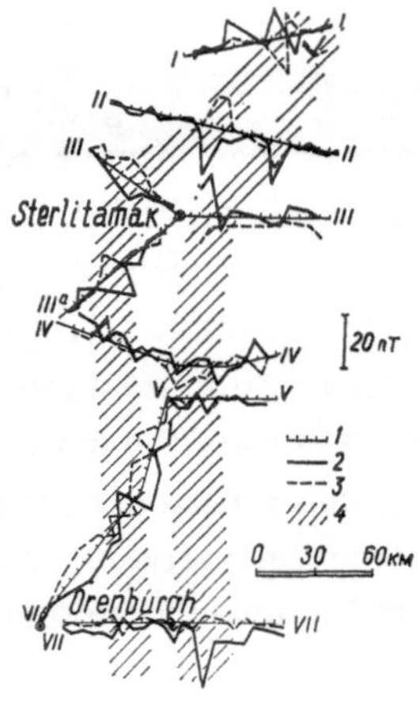

Fig. 3. Bashkir secular variation anomaly. 1, survey points; 2, curves of differences ΔT for 1974–1973; 3, curves of differences ΔT for 1975–1974; 4, anomalous zones.

to the Bashkir one. Though this anomaly is seen in two profiles, it is still only poorly defined. It is therefore not analyzed in the present work.

The Butka anomaly is seen over 200 km of a territory under exploration (Fig. 1), and coincides with the electrical conductivity anomaly, elongated along the Talitza megasynclinorium of the Zaurals uplift (VISHNEV *et al.*, 1972).

Thus the zone of the folded Urals is separated from the contiguous section of the Russian platform and the Western Siberian plate by regions which exhibit secular variation anomalies of a geomagnetic field T of a single type.

3. Geodynamic Data of the Urals

3.1 Results of the stress measurements in the zone of the folded Urals

To determine the character of tectonic forces acting in the zone of the folded Urals, measurements of elastic stresses in the rocks were made in mines at twelve sites (Fig. 4 and Table 1). The results allow the definition of tectonically active forces. The tectonic forces are compressional. They act preferentially in a horizontal plane, the greatest compressive stress is oriented perpendicular to the strike of Urals structures (BULASHEVITCH *et al.*, 1976).

The elastic stresses field is non-uniform within the limits of explored depths

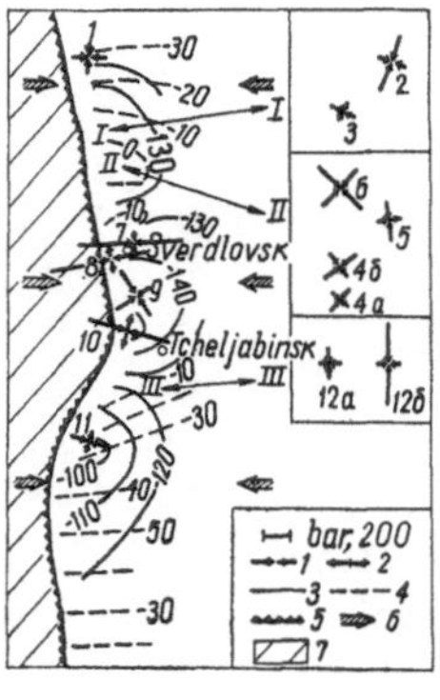

Fig. 4. The stress-pattern distribution in upper part of the Urals lithosphere according to measurements in situ and model investigation data. 1–2, stress and strain according to measurements in the mining output; 3, isostate of σ_a stresses (model); 4, isostate of σ_b stresses (model); 5, joining zone of the miogeosynclinal and eugeosynclinal Urals zone; 6, direction of external efforts (model): 7, rigid part of model.

Table 1. Characteristics of rock stress conditions in the upper part of the crust at the Urals according to the mine output measurement data.

No.	Name of measurement points	Depth (m)	Azimuth (°)	Rock stresses in the massif (bar)			Tension calculated by rock's weight (bar)	
				Horizontal		Vertical	Vertical	Horizontal
1.	North-Peschyansky	300	100	−135	−7	−10	−84	−28
2.	Goroblagodatsky	170	98	−178	−238	−146	−48	−16
3.	Valuevsky	120	118	−47	−19	−27	−34	−11
4.	Vysokogorsky							
	a)	370	130	−65	−95	−106	−100	−33
	b)	326	54	−140	−97	−105	−90	−30
5.	Lebyazjinsky	330	74	−44	−76	−13	−92	−31
6.	Estuninsky	308	135	−205	−215	−177	−84	−28
7.	Beryezovsky	300	90	−425	−74	−212	−84	−28
8.	Degtyarsky	430	83	−490	+90	+60	−127	−42
9.	Karabashsky	700	100	−335	+132	−160	−196	−65
10.	Vyshnevogorsky	100	60	−177	−183	−125	−28	−9
11.	Mindyaksky	247	125	−92	−41	−11	−69	−23
12.	Kochkarsky							
	a) Mine 116	192	90	−56	−107	−20	−56	−18
	b) Mine "Center"	295	90	−70	−313	−95	−84	−28

Mark + (plus) stands for tension, mark − (minus) stands for compression.

(up to 700 m), with a value reading 500 bars. The greatest concentration of elastic stresses is observed at external sections of convex geological structures. Bashkir SVA is situated in just such a section.

3.2 Laboratory experiments with models

The Manchazh anomaly is situated at the Ufa section of the Russian platform, for which no direct measurement of the stress in situ was available. To estimate the stressed state, laboratory modelling was used. To check the validity of the exploration of stress distribution using photoelastic models, the variation of stress in the Urals and the Russian platform was simulated and checked against observations. As a first variant the border between the Urals and the Russian platform was drawn at the Eastern margin of the Preduralsky trough, and as a second variant 100 km to the East of the Western border of Tagil-Magnitogorsk structural line. In both models under consideration, the pattern of the stress distribution proved to be similar and roughly latitudinal compressional forces were present. Beyond the platform projection, the zone of shearing off and sublatitudinal compressional stresses concentration is observed. This corresponds to the zone of relatively high stress between Sverdlovsk and Tcheljabinsk (measurements points 7, 8, 9, 10; Fig. 4). Meridional compression is essentially decreased and even tensional stresses appear. Inside the platform projection, owing to the reaction of the parted blocks of the earth's crust, stresses in a horizontal plane are probably a little rotated, the least compression stress is oriented vertically.

As the predicted tectonic forces in the Sverdlovsk and Tcheljabinsk regions have a similar pattern to those measured, it was reasonable to assume that those may be determined approximately from the model of the Ufa projection of the Russian platform.

3.3 Elastic stresses behavior in time

The presence of tectonically active forces, operating for a long time, combined with heterogeneous stress fields, suggest that in several sections zones of failure may arise. Consequently, the stress fields may change in time. It is clear that the development of the weakened sections of the earth's crust can take place without intensive accumulation of the elastic stresses, because of the difference between long-term rigidity and an experimentally determinable value (ZUBKOV and MYACHKIN, 1975). The essence of the phenomenon is that material fails under the influence of long-term application of constant stress at much lower stress than the rigidity measured for rapid stressing. The data on the seismic activity at the Urals reveals the presence of the temporal stress field variations. However, until recently, no direct measurements of stresses variations in time had been made in the Urals. The first attempt to make such measurements was in 1975. Photoelastic pressure sensors with a sensitivity of 0.2 bar were established at the point 7 (Fig. 4), which is situated in the zone of stress concentration. After the measurements, repeated many times during a year and a half, the change of elastic stresses was found to be less than the background ± 3 bars in the section investigated.

3.4 Laboratory investigations of the rock magnetic properties dependence on pressure

The possible observation of tectonic stresses led to the measurement of rock magnetic property variations as a result of hydrostatic and uniaxial compression. Similar experiments have not been attempted in the past. Two extreme cases were modelled: uniaxial and hydrostatic compressions. To our mind, hydrostatic pressure possesses an essential advantage, as it allows the laboratory conditions to simulate pressures near to those which are supposed at a magnetoactive layer of the lithosphere.

Thermoremanent magnetization I_{rt} is the strongest and the most stable of all the remanent magnetization in a given field. Let us see if two processes, the change of

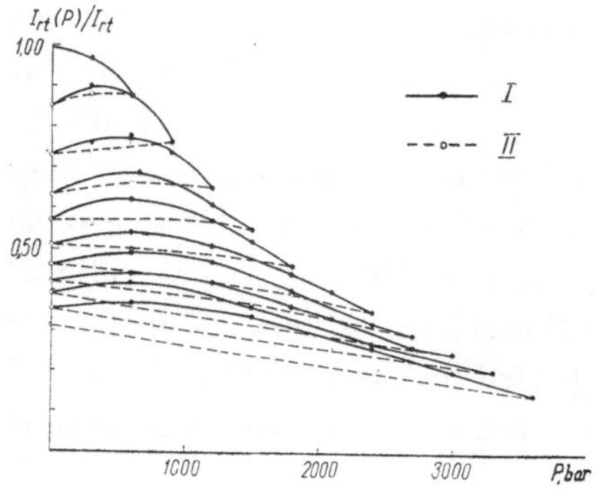

Fig. 5. Relative changes of thermo-remanent magnetization of the magnetite ore sample under the cyclic variation of hydrostatic pressure with ever increasing amplitude. I, rise of pressure; II, reduction of pressure.

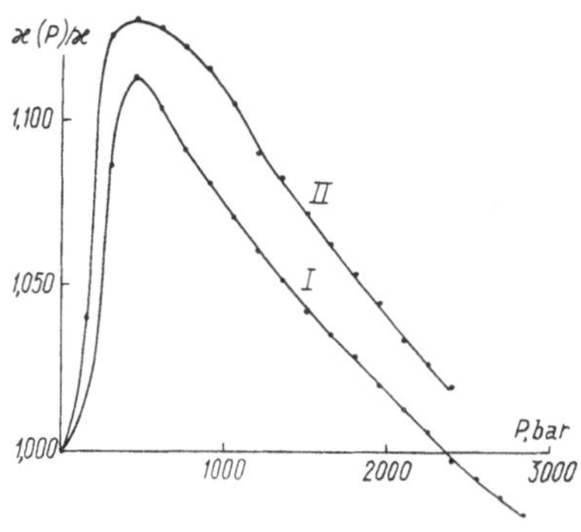

Fig. 6. Relative changes of magnetic susceptibility of the magnetite ore sample on the hydrostatic pressure increase. Curve I, the first loading; curve II, the second loading.

a thermoremanent magnetization and/or magnetic susceptibility, can provide the variations of the field at the Manchazh surface anomaly of up to 2–3 nT/year. The upper surface of the source lies here at a depth of not less than 4–5 km (BUGAYLO *et al.*, 1976) which corresponds to a pressure of a about 2 kbars. The change of I_{rt} upon stress cycling is presented in Fig. 5 (solid curves are load; dotted lines, discharging) (IVANOV *et al.*, 1973). Changes of magnetic susceptibility measured in variable field during 2 cycles of stressing, carried over with an interval of 30 days between cycles, are shown in Fig. 6 (numbers of curves correspond to the numbers of cycles; sample's discharging is not shown). One can see that in the region of interest, both processes are of the same type. If the pressure exceeds 2 kbars, κ and I_{rt} decrease with pressure. Our experiments show that these changes are reversible for κ from the first cycle, for I_{rt} after some cycles. We investigated magnetite ore samples with magnetic susceptibility from 0.03 to 0.40 units CGSM, this stress sensitivity of the rocks varies from 0.0001 units CGSM/bar to 0.00001 units CGSM/bar.

4. Discussion of the Results

Having utilized the results of field and laboratory experiments, we develop several ideas on the nature of the observed secular variation anomalies. We may try to interpret the Manchazh anomaly (the anomaly of the first type) from the point of view of tectonomagnetic effect, by estimating the value of yearly changes of the stress, which can give the observed increments of a field. If the Manchazh anomaly is caused by the rocks with κ about 0.01 units CGSM (BUGAYLO *et al.*, 1976), then from the calculation of the maximum sensitivity 0.0001 units CGSM/bar and the pressure change for 10 bars, we obtain a field variation at the surface of 1 nT (anomalous field is of 1,200 nT). Variation may be higher, if the magnetic moment due to viscous remanent magnetization, which is more sensitive to the pressure (BEZUGLAIA *et al.*, 1973) makes a contribution.

The absence of any spatial linkage of the Bashkir and Butka SVA with static magnetic field anomalies, and the large magnitude of as much as 20 nT/year of anomalous changes of the field T, strongly suggests that they were not of a piezomagnetic nature. Their disposition in the zones of electrical conductivity anomalies leads to the idea that electrokinetic events cause these anomalies. In the zone of the

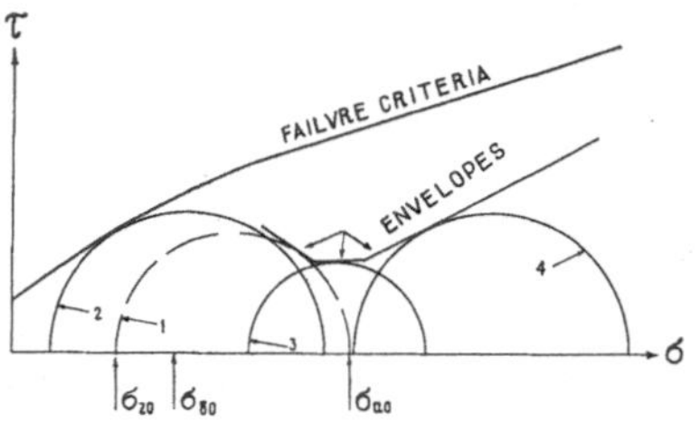

Fig. 7. Mohr diagrams for initial effective stresses characteristic of thrust faulting from MARTIN (1975). Curve 1, initial conditions; curve 2, failure conditions with increased fluid pressure; curve 3, decreased fluid pressure; and curve 4, further decreased fluid pressure.

Bashkir SVA the ratio of effective stresses σ_{a0}, σ_{b0}, σ_{z0} characterizes the condition of thrust dislocation development, the least stable to the influence of pore waters. The connection of main effective stresses σ_a, σ_b, σ_z with the change of pore pressure P is obtained by MARTIN (1975) in the following expressions:

$$\sigma_a = \sigma_{a0} - [\gamma/(1-\gamma)] \cdot \alpha \cdot \Delta \cdot P$$
$$\sigma_b = \sigma_{b0} - [\gamma/(1-\gamma)] \cdot \alpha \cdot \Delta \cdot P$$
$$\sigma_z = \bar{\rho} \cdot g \cdot H - \alpha \cdot P_0 - \alpha \cdot \Delta \cdot P$$

where $\bar{\rho}$ is average density of the rocks, H is the depth, γ is Poisson's coefficient, α is elastic constant (NUR and BYERLEE, 1971), σ_a, σ_b, σ_z—maximum horizontal, minimum horizontal and the vertical effective stresses. From those equations it follows that the ratios $\Delta\sigma_a/\Delta\sigma_z$ and $\Delta\sigma_b/\Delta\sigma_z$ change within the limits $0.11 \div 0.55$ with the change γ $0.1 \div 0.35$. From Mohr's diagrams (Fig. 7) we can see that with the ratio of initial effective stresses $|\sigma_{a0}| > |\sigma_{b0}| > |\sigma_{z0}|$ the conditions of rock failure may easily be achieved. Spatial redistribution of pore fluid pressure and the change of their mineralization (OSIKA *et al.*, 1977) will be reflected in their electric conductivity and could lead to electrokinetic phenomena in thrust zones such as the region of the Bashkir secular variation anomaly. Redistribution of the electric current between enclosing rocks and conductive zone leads to the anomalous changes of the magnetic field in time. Thus, the geomagnetic field variations probably reflect geodynamical conditions of active thrusts, responsible for the formation of modern structures of the Urals, though they may be partially caused by the events of electromagnetic induction in the conductive zones, which spatially coincide with the secular variation anomalies. The active thrust zone in Predurals coincides with the sections of maximal for Urals horizontal gradients of the gravity isostatic anomalies and modern vertical motions.

5. Conclusion

This paper is a preliminary attempt to link the results of a highly accurate geomagnetic survey, observed stress on the geodynamics of the Urals and the laboratory experimental data together. The contemporary state of those measurements allows us to make only qualitative conclusions.

The main conclusion of the present work is that although the current level of our knowledge does not permit an understanding of the SVA nature, highly accurate geomagnetic survey is one of the most sensitive methods for studying of modern processes in the earth's crust.

The method of direct stress measurements in rocks still gives a static pattern only (within the limits of the measurement error ± 3 bars). This circumstance does not allow us to fix small stress changes. By the energy and periodicity of tectonic earthquakes, we may judge indirectly and, unfortunately, in a qualitative way only about the stress changes.

If it proves possible to establish the connection between anomalous field variations and stress changes unequivocally, a simple method to measure stresses in situ is available. To solve that problem it is necessary, first, to find the tectonic processes by quantitative investigation, and secondly, it is necessary to model the supposed processes in laboratory conditions.

REFERENCES

BARSUKOV, O.M., L.S. BEZUGLAIA, V.N. VADKOVSKY, and Yu.P. SKOVORODKIN, About a nature of one secular variation anomaly of the earth's magnetic field, *Izv. Acad. Sci. USSR, Fizika Zemli*, **9**, 85–90, 1968.

BEZUGLAIA, L.S., S.H. MAKSUDOV, and Yu.P. SKOVORODKIN, Viscous magnetization of the rocks under pressure, *Izv. Acad. Sci. USSR, Fizika Zemli*, **4**, 110–114, 1973.

BUGAYLO, V.A., V.S. DRUZHININ, G.G. ORLOV, and L.F. RIBALKA, On geological structure of the Manchazh regional anomaly nature, in *Structure and Development of the Earth's Crust and Ore Fields of the Urals due to Geophysical Data*, 29–35, Ural Scientific Center, U.S.S.R. Academy of Sciences, Sverdlovsk, 1976.

BULASHEVITCH, Yu.P. and V.A. SHAPIRO, Anomaly of the geomagnetic field secular variation of the Arti observatory, *Geomagnetism y Aeronomia*, **15**, 382–384, 1975.

BULASHEVITCH, Yu.P., A.L. ALEYNIKOV, O.V. BELLAVIN, V.A. BUGAYLO, N.I. KHALEVIN, V.A. SHAPIRO, I.F. TAVRIN, N.P. VLOKH, and A.V. ZUBKOV, Geodynamics of the Urals, *Tectonophysics*, **35**, 15–26, 1976.

IVANOV, N.A., A.A. NULMAN, and V.A. SHAPIRO, Reversible and irreversible changes in TRM under hydrostatic compression, in *Constant Geomagnetic Field, Paleomagnetism and Rock Magnetism*, edited by G.N. Petrova, pp. 148–150, Naukova dumka, Kiev, 1973.

MARTIN, J.C., The effect of fluid pressure on effective stresses and induced faulting, *J. Geophys. Res.*, **80**, 3783–3785, 1975.

MIZUTANI, H. and T. ISHIDO, A new interpretation of magnetic field variation associated with the Matsushiro earthquake, *J. Geomag. Geoelectr.*, **28**, 179–188, 1976.

NAGATA, T., *Rock Magnetism*, pp. 258–288, Maruzen, Tokyo, 1961.

NUR, A. and J.D. BYERLEE, An exact effective stress law for elastic deformation of rock with fluids, *J. Geophys. Res.*, **76**, 6414–6419, 1971.

OSIKA, D.G., A.B. MEGAEV, T.S. YANKOVSKAYA, and O.A. SAIDOV, Hydrogeochemical anomalies, precursory of earthquakes as reflection of the conditions of its focal zone formation, *Dokl. Acad. Sci. USSR*, **233**, 74–78, 1977.

SHAPIRO, V.A., Seismomagnetic effect, *Izv. Acad. Sci. USSR, Fizika Zemli*, **8**, 61–74, 1968.

SHAPIRO, V.A. and N.A. IVANOV, Investigations of secular variation anomalies in the Middle Ural region, *J. Geophys.*, **40**, 435–438, 1974.

SHAPIRO, V.A., Local secular variation anomalies and earthquake prediction problem, in *Earthquake Forerunners Searching*, edited by E.F. Savarensky, pp. 200–207, Fan Publishers, Uzbek SSR, Tashkent, 1976.

SHAPIRO, V.A. and V.A. PYANKOV, Current anomaly of the geomagnetic field *T* secular variation in Bashkiria, *Geomagnetism y Aeronomia*, **16**, 943–945, 1976.

SOBOLEV, G.A., V.N. BOGAEVSKY, R.A. LEMENTUEVA, N.I. MIGUNOV, and A.A. KHROMOV, Investigation of mechanoelectric events in the seismoactive region, in *Physics of Earthquake Focus*, edited by M.A. Sadovsky, pp. 184–223, Nauka, Moskow, 1975.

STACEY, F.D., K.G. BARR, and G.R. ROBSON, The volcano-magnetic effect, *Pure Appl. Geophys.*, **62**, 96–108, 1965.

TRUKHIN, V.I., *Introduction in Rock Magnetism*, pp. 233–238, MGU, Moskow, 1973.

VISHNEV, V.S., A.G. KRASNOBAEVA, V.V. KORMILTSEV, and A.E. RITSK, Magnetic deep sounding in the Middle Ural region, in *Memories of Institute of Geophysics*, pp. 26–30, Ural Scientific Center, U.S.S.R. Academy of Sciences, Sverdlovsk, 1972.

ZUBKOV, S.I. and V.I. MYACHKIN, About permanent rigidity of rock masses of earthquake focal zones, in *Physics of Earthquake Focus*, pp. 57–67, Nauka, Moskow, 1975.

Geomagnetic Investigations in the Seismoactive Regions of Middle Asia

V.A. SHAPIRO,* A.N. PUSHKOV,** K.N. ABDULLABEKOV,***
E.B. BERDALIEV,*** and M.YU. MUMINOV***

**Institute of Geophysics, Academy of Sciences, Sverdlovsk, U.S.S.R.*
***Institute of Earth Magnetism, Ionosphere and Radio Wave Propagation, Academy of Sciences, Moscow, U.S.S.R.*
****Institute of Seismology, Uzbek Academy of Sciences, Tashkent, U.S.S.R.*

(Received September 1, 1977)

Investigations of local geomagnetic field changes have been carried out in Middle Asia in the following regions: Tashkent, including Charvak reservoir, Fergana and Kyzyl-Kum. Since 1968, a variety of total field features have been observed there, including: (a) slow changes at separate stations, which may be explainable by compression or tension of individual earth's crustal blocks; (b) a variety of field changes with amplitudes of a few gammas and periods of 0.5–2 years; (c) variations, which may be due to different conductivity of rocks; (d) anomalous changes in the fracture zones; (e) anomalous variations in the Charvak region, connected with reservoir filling. The amplitudes of these variations are 3–5 nT/year at Fergana, 8–9 nT/year at Kyzyl-Kum, 20–25 nT/year at Tashkent regions and up to 15–25 nT/year at Charvak reservoir area. Sometimes these anomalous changes are correlated with seismic activity (for example, Tashkent 1968 earthquakes).

1. Introduction

At the present time, investigations of secular variation anomalies of the geomagnetic field in the Middle Asia are underway in the geodynamical polygons of Tashkent, Fergana and Kyzyl-Kum, and in the area of the water reservoir of Charvak (Fig. 1). Those polygons are situated in the areas of different seismotectonic conditions. Fergana polygon is in an active orogenic area of the Central Tyan-Shan. Kyzyl-Kum polygon is situated within a platform, characterized by intensive recent fracturing. The Tashkent polygon is in a transition zone between the orogene and platform. Geomagnetic field variations, caused by the natural processes in the earth's crust were observed at all the polygons. In the area of the water reservoir of Charvak geomagnetic field total force secular variations, induced by changeing pressures in the reservoir, have been studied.

2. Tashkent Polygon

Geomagnetic studies in the Tashkent polygon were started in 1968. Measurements were made with proton precession magnetometers by the method of repeated

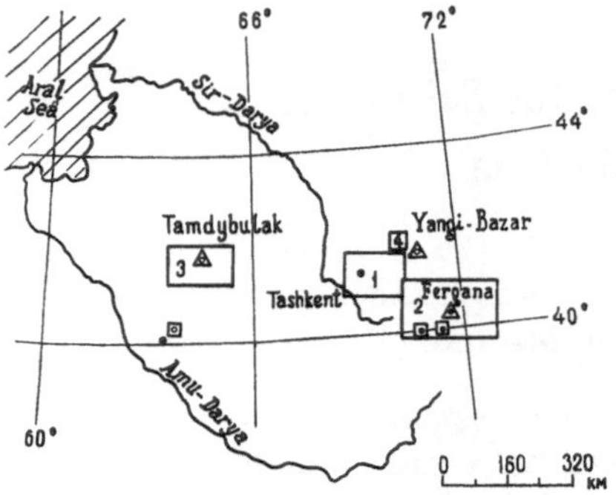

Fig. 1. Sketch map of Middle Asian geomagnetic polygons.

route surveys. The distance between observations points was 4–5 km at the polygon and 1–2 km in the region of the water reservoir. Measurements in the Tashkent polygon were made on three routes, lying along the rupture zone and a series of faults (Abdullabekov and Maksudov, 1975). The frequency of the measurements was 3–4 cycles a year. The mean square error of measurements was ± 2 nT. Observations were related to the total force field T of the Yangi-Bazar magnetic observatory. The distance from the magnetic observatory to the measurement points does not exceed 70 km. In 1971–1972 anomalous changes of the geomagnetic field total force with 20–25 nT intensity were observed on a route lying along the zone of rupture. These changes appear to be associated with an earthquake of magnitude M=4–4.5 which took place near the route. The following year measurements on the routes were continued. However, anomalous changes of the field, exceeding by a factor of 3 the mean square error, were not seen. This is consistent with the seismological data as there were no earthquakes with magnitude M over 2.5 near the observation points during 1973–1977.

3. Kyzyl-Kum Polygon

Measurements in the Kyzyl-Kum polygon were made at the network of the points established in 1974 (Fig. 2). Measurements of total field were made synchronously at the points of reference. The repetition frequency was one cycle a year. Measurements during the cycles were repeated at every point twice in different phases of daily variations of total force. The distance between the reference and the observation points does not exceed 80 km. Routes cross active faults and sections with intense fracturing and different structural blocks. The anomaly of the stationary field at the routes reached some hundred nano Teslas. The depth of the Paleozoic basement is 0–500 m. Measurements at the points were made in 1974–1975 with a proton magnetometer IIM-5, in 1976–1977 with proton magnetometers G-816 of the 'Geometrics' firm and T-MII of the Institute of Geophysics, Urals Scientific Center, USSR Academy of Sciences (Maksimovskikh and Shapiro, 1976). The mean square error of the measurements in 1974 was ± 1.3 nT, in 1975 ± 1.8 nT, and in 1976–1977 ± 0.5 nT. Changes of the field on the routes for 1974–1975, 1974–1976, 1976–1977 are presented in Fig. 3.

As is seen from the diagrams, the greatest changes of the field were observed in 1974–1975 and 1975–1976. The smallest changes of the field were observed in 1976–1977. This is most clearly seen in the histograms of field variations. Histograms showed least variation in 1974–1975 and 1974–1976 (Fig. 4a, b) and most

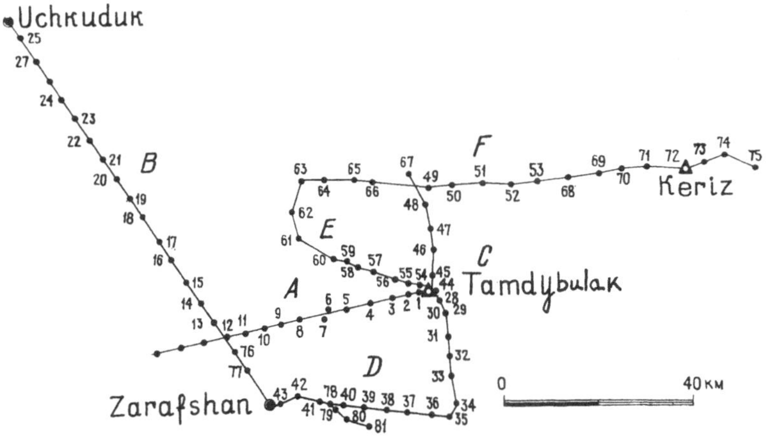

Fig. 2. Sketch map of Kyzyl-Kum geomagnetic polygon. Dots and digits symbolize the survey points; triangles, point of reference. A, B, C, D, E, F: survey profiles.

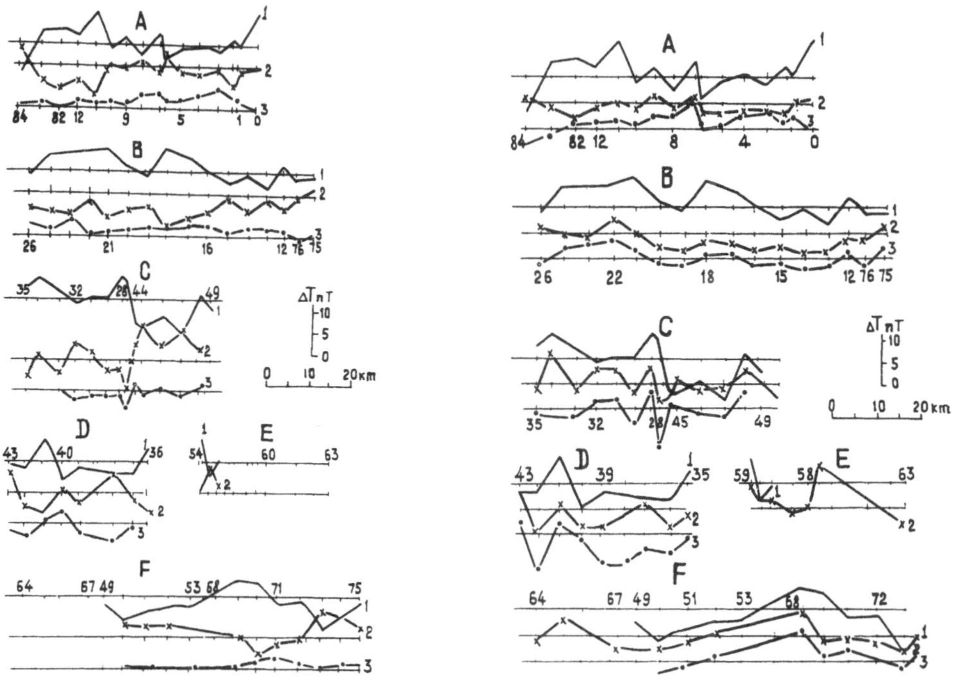

(a) (b)

Fig. 3. Changes of the geomagnetic field T along the profiles of Kyzyl-Kum polygon. (a) 1, ΔT difference (1974–1975); 2, ΔT difference (1974–1976); 3, ΔT difference (1974–1977). (b) 1, ΔT difference (1974–1975); 2, ΔT difference (1975–1976); 3, ΔT difference (1976–1977). A, B, C, D, E, F: survey profiles (Fig. 2).

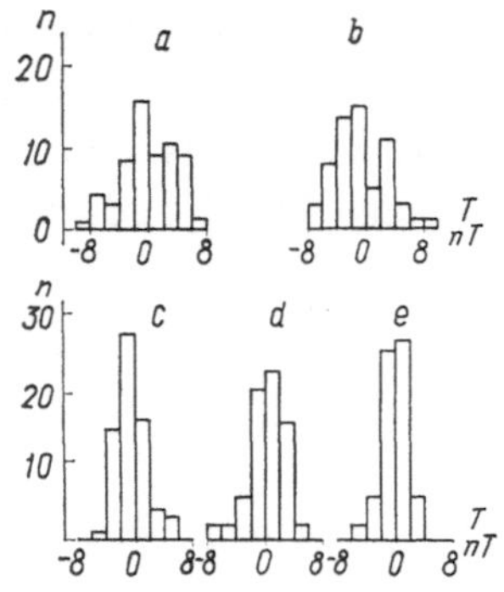

Fig. 4. Histograms for total force T distributions during 1974–1975 (a), 1975–1976 (b), 1974–1976 (c), 1974–1977 (d), 1976–1977 (e). n: number of points.

variation in 1976–1977 (Fig. 4c, d, e). This suggests to us that the changes of the field in that polygon in 1975 were probably anomalous. Observations of 1975 were made 5–6 months before the earthquakes of the 8th of April and 17th of May, 1976 in Gazly with the magnitude $M=7$–7.5. The next observations were made in October 1976. Probably the whole territory of the polygon was in a stressed state before the earthquakes. According to the modern opinion the region of elastic stress accumulation occupies much more territory than was considered earlier. Thus before the Gazly earthquakes, the region of stress accumulation with the deformation value 10^{-7} may have occupied territory with a radius of 650–700 km.

The points of Kyzyl-Kum polygon were at distances of 180–250 km from the epicenter of the Gazly earthquakes. Therefore the territory of the polygon could exist in a stressed state. The distribution of fracturing which had taken place favors this view. In 1974–1975 intensive processes of fracturing were seen at some sections of the polygon. After the earthquake these processes sharply decreased. Probably, the Gazly earthquakes and fracturings were caused by the same source.

Our statement is still an assumption, and not a proven fact. However, we can add that the average square error had its greatest value ± 1.8 nT in the anomalous (1975) year. Also supporting this supposition is the fact that the maximal change of the field 8 nT, discovered at the point of reference of Tamdybulak is in the region of fracturing. This change was seen in 10-day averages (one reading every 10 min) in 1974–1975. Total field variations in Tamdybulak were compared with those of the magnetic observatories Yangi-Bazar, Vannovskaya, and Novokazalinsk.

4. Fergana Polygon

Measurements at Fergana polygon were made for 1972–1973 by the closed route with a total length of 650 km (Fig. 5). The routes passed through seismically active zones, enveloping abyssal fractures, epicenters of strong earthquakes, gradients of modern and the recent motions of the earth's crust and gradients of geophysical fields. Until 1975 the Yangi-Bazar observatory was used as a point of reference for the Fergana polygon. The distance between the observatory and the southeastern part of the polygon is 350 km. The average square accuracy of the measurements

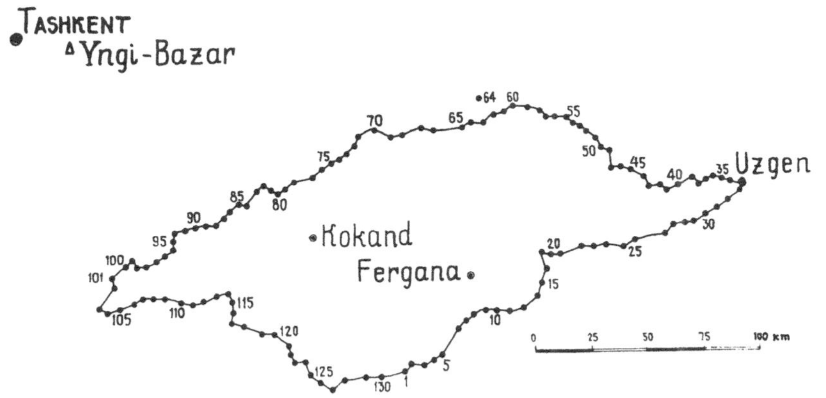

Fig. 5. Sketch map of Fergana geomagnetic polygon. Dots and digits symbolize survey points.

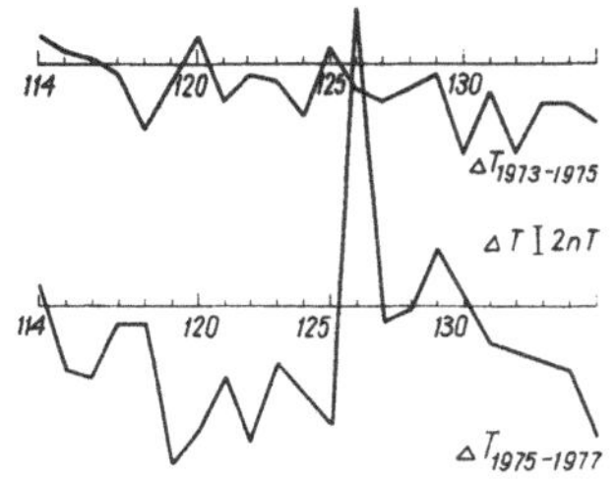

Fig. 6. Changes of the geomagnetic field total force T along the profile Kanibadam-Vuadil (points 114–135).

at that polygon was $\pm$3–4 nT in 1975. Measurement errors because of incorrect calculation of the variations at separate points could be as large as 10 nT. In that connection the results of the measurements of 1975 are not discussed in detail. In 1975 after using a synchronous method of measurement with the magnetometer T-MII (Maksimovskikh and Shapiro, 1976) the average square error of measurements decreased significantly.

Changes of the field in the route Kanibadam-Vuadil (points 114–130 and 1–15) appear to be anomalous. Here, in the direct neighborhood of the route (5–6 km to the northeast) a strong earthquake with a magnitude $M=5.75$ took place on the 31st of January, 1977. The depth of the hypocenter was about 15 km. Changes of the field at the route for 1973–1975 and 1975–1977 are presented in Fig. 6. Unfortunately, the latest investigations before the earthquake were made in that part of the route in 1975. The repeated measurements were made 10 days after the earthquake. Large field changes in 1973–1975 were not observed. In the period 1975–1977 we can observe, in the area of the 125th point, sharp changes of the field. The maximum change along the route reached 29 nT. Besides that, changes of the field level between points 114–125, 125–130 and 1–5 are marked. The average change is about 15 nT. Probably, the observed anomalous variations of the geomagnetic field are due to the redistribution of elastic stresses after the strong earthquake. We suggest that the position of the maximum change indicates the zone of maximum stressing associated with the epicenter of the earthquake. It is worth remarking that the epicenter of the next strong earthquake with the $M=5.5$ of the 3rd of June, 1977 shifted in that direction to a distance of 30–40 km.

5. Water Reservoir of Charvak

Geomagnetic field variations were observed in detail in a region in which a water reservoir was being filled. A network of points was located in 1973 around the water reservoir of Charvak. The distance between the observation points was 1–2 km and they were removed from the water reservoir bank by some hundred meters of up to 10 km. The measurements were repeated once every 1.5–2 months for winter months and 1–2 times a month in other periods of the year. Yangi-Bazar observatory was used as the point of reference. The distance between the observatory and the field points is 50–60 km. Anomalous variations of the field at the majority of the points were correlated with the change of the water level in the reservoir. Maximum changes in separate points were as high as 15–25 nT. At points near to the reservoir, the intensity of the changes was greater. Average changes of the field at 35 points for 1974–1976 is presented in Fig. 7. The value of the average change achieves 10 nT in 1975–1976. The intensity of the T anomaly depends on the volume of water—the greater quantity of the water in the reservoir, the greater was the amplitude of the variations. In spite of the obvious linkage between the geomagnetic field changes and the water reservoir filling, the nature of these anomaly variations cannot be explained simply, for the observed anomalies could not be caused by excessive pressure only, which rises on filling, but by anomalous electroconductivity, appearing while moistening the soil around the water reservoir. Further research will evidently allow us to speak more definitely about the nature of such anomalous variations of the geomagnetic field.

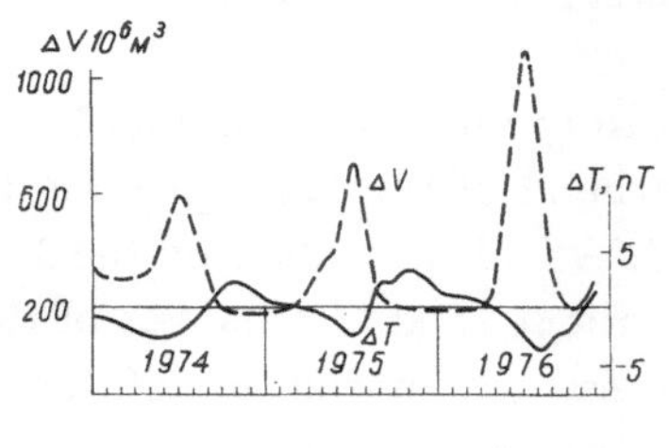

Fig. 7. Variations of the geomagnetic field total force T as a result of the changes of water volume in Charvak reservoir.

6. Conclusions

Having analyzed the results of the geomagnetic variation in geodynamical polygons situated in seismoactive regions of the Middle Asia, we conclude:

1) In all the polygons—namely, Tashkent, Kyzyl-Kum, Fergana—we have observed anomalous variations of geomagnetic field total force T with magnitudes up to 25 nT. In some cases such variations appear to be correlated with earthquakes.

2) At all the enumerated polygons the degree of anomalous behavior of the secular variation field was qualitatively correlated to the seismic activity.

3) At Kyzyl-Kum polygon the process of active fracturing terminates after the strong earthquakes of the 8th of April and 17th of May, 1976.

4) At the water reservoir of Charvak distinct correlation of an anomalous variation of the field with the level of reservoir filling has been observed.

The present results indicate that on the basis of the geomagnetic research at the geodynamical polygons, widening of the area of observation and improvement of the observation procedure will enable us to understand fully the tectonomagnetic processes in this area.

REFERENCES

ABDULLABEKOV, K.N. and S.Kh. MAKSUDOV, *Geomagnetic Field Variations in Seismoactive Regions*, pp. 34–55, FAN Publishers, Tashkent, 1975.

MAKSIMOVSKIKH, S.J. and V.A. SHAPIRO, Portable proton precession magnetometer of high accuracy T-MII, *Geomagnetism y Aeronomia*, **XVI**, 389–391, 1976.

Local Magnetic Field Variations and Stress Changes Near a Slip Discontinuity on the San Andreas Fault

M.J.S. JOHNSTON

U.S. Geological Survey, 345 Middlefield Road, Menlo Park, California 94025, U.S.A.

(Received November 25, 1977)

Data from an array of proton magnetometers in central California indicate that a systematic decrease in magnetic field of about 2γ in 5 years has occurred in a localized region near Anzar, California, just north of the creeping section of the San Andreas fault. This field change has most likely resulted from changes in crustal stress in this region, although an unknown second-order effect of secular variation cannot be excluded as a alternate explanation. Tectonomagnetic models have been developed using dislocation modeling of slip on a finite section of fault. Assuming a fault geometry and rock magnetization, these models relate changes in stress, fault slip, and fault geometry to surface magnetic field anomalies. A large-scale anomaly, opposite in sense to that observed but of similar amplitude, would be expected to have accumulated in this area during the past 70 years. A localized 5-bar decrease in shear strain on the fault resulting from about 2 cm of slip on a 0.25-km square patch at a depth of 1 km beneath the surface trace of the fault opposite the magnetometer could explain the observed data and still be compatible with the geodetic strain measurements in the area. Other models of limited local slip are equally possible. The occurrence of a moderate magnitude earthquake in this region will allow comparison of stress changes estimated by different techniques.

1. Introduction

The magnitude of crustal stress changes accompanying faulting is uncertain. On one hand direct laboratory measurements (STESKY and BRACE, 1973) and deep mine fracture experiments in intact rock (SPOTTISWOOD and MCGARR, 1975) indicate that the shear stress at about 10 km is 1 to 2 kbars and that stress changes of up to 1 kbar should therefore accompany faulting. On the other hand an upper limit on the average shear stress on the fault of a few hundred bars is indicated by the absence of a detectable heat flow anomaly near the San Andreas fault (BRUNE *et al.*, 1969; LACHENBRUCH and SASS, 1973). The mean displacement-to-length ratios for earthquakes that rupture the surface also indicate an average change in stress of about 100 bars (CHINNERY and PETRAK, 1968). Furthermore, BRUNE (1970) and others have argued from theoretical seismic source models for seismic shear stress changes (stress drops) that are typically less than 100 bars. Comminution and chemical changes probably also contribute to minor changes in stress levels on active faults but are not sufficient to explain this challenging paradox.

Independent estimates of changes in stress distribution around active faults are possible as a consequence of the stress sensitivity of the magnetization of crustal rocks (STACEY, 1964). These estimates are generally poorly constrained since the distribution of magnetization and its stress sensitivity are poorly known. Within these uncertainties, however, meaningful order-of-magnitude stress estimates can be determined. These can be related to the paradox discussed above. This paper primarily concerns the use of a tectonomagnetic model to attempt a stress estimate at the location of a particular slip discontinuity on the San Andreas fault where a change in local magnetic field has been observed.

2. Local Magnetic Anomalies

The history of high-quality magnetic measurements along active faults in California is extremely short (JOHNSTON *et al.*, 1976). It is interesting, but possibly coincidental, that even with this short history, the regions where significant local anomalies have been reported are at or near the ends of the recent major ruptures ($M > 8$) within the fault system. These earthquakes are shown in Fig. 1, together with all earthquakes $M \geqq 5$ since 1900 and the amplitude and error estimates of the

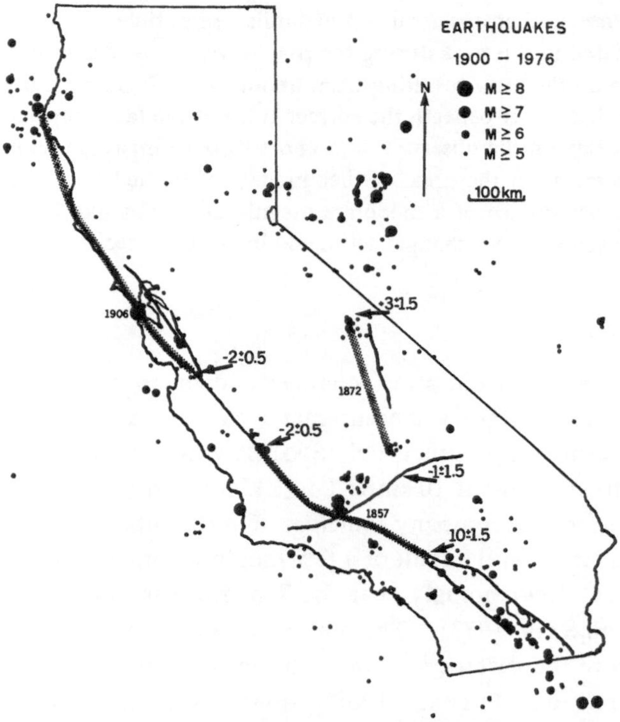

Fig. 1. Major recent fault ruptures (cross-hatched) for earthquakes $M > 8$ and the locations of earthquakes with $M \geqq 5$ (dots) in California and western Nevada since 1900 from TOPPOZADA *et al.* (1976), BOLT and MILLER (1975), HILEMAN *et al.* (1973), and FRIEDMAN *et al.* (1976). Included also are the amplitude and error estimates of magnetic anomalies that have occurred since 1973.

magnetic anomalies that have occurred since 1973. Details of most of the magnetic data can be found in JOHNSTON *et al.* (1976), WILLIAMS and JOHNSTON (1976), and SMITH and JOHNSTON (1976).

The data that form the subject of this study have been recorded just north of San Juan Bautista at the south end of the 1906 earthquake rupture. The locations of continuously recording magnetometers in this region are shown in Fig. 2. All magnetometers are synchronized. The sensitivity and stability of each is 0.25γ and the sampling rate is once per minute (SMITH and JOHNSTON, 1976). The data from each adjacent station are differenced to isolate changes of local origin and to reduce effects of ionospheric and magnetospheric origin.

The magnetizations of the rocks in this area can exceed 10^{-3} emu (in some places 10^{-2} emu) and thus provide a sensitive stress transducer in this region. Similar clustering of rocks with high magnetizations occurs also at the ends of the 1857 earthquake ($M>8$) rupture in southern California and the 1872 earthquake ($M>8$) rupture in Owens Valley.

Space-time plots for the period 1974 to 1976 of differential magnetic fields across the region from where the fault is locked to where the fault is slipping are shown in

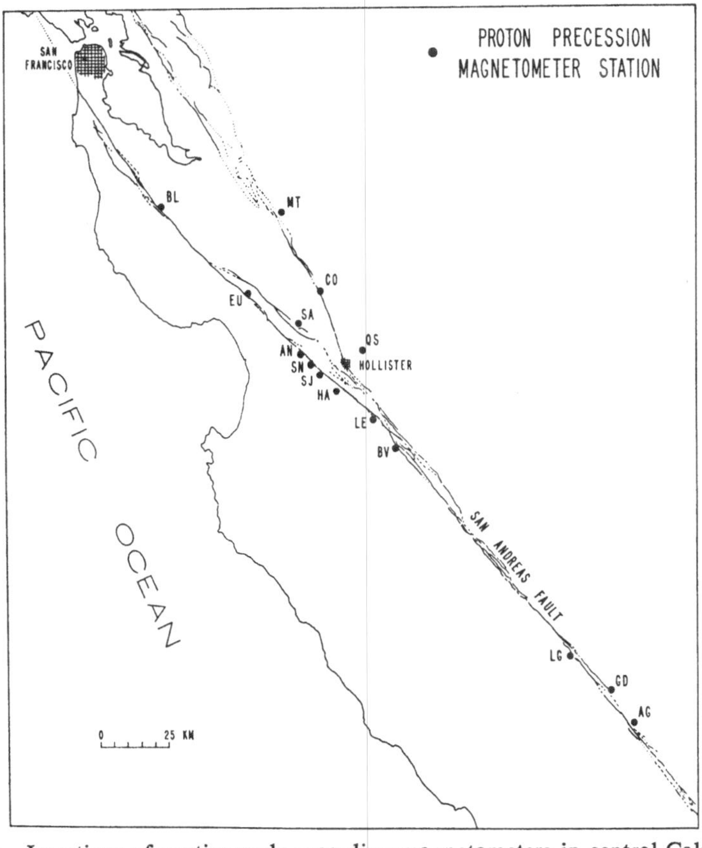

Fig. 2. Locations of continuously recording magnetometers in central California presently telemetered to Menlo Park just south of San Francisco.

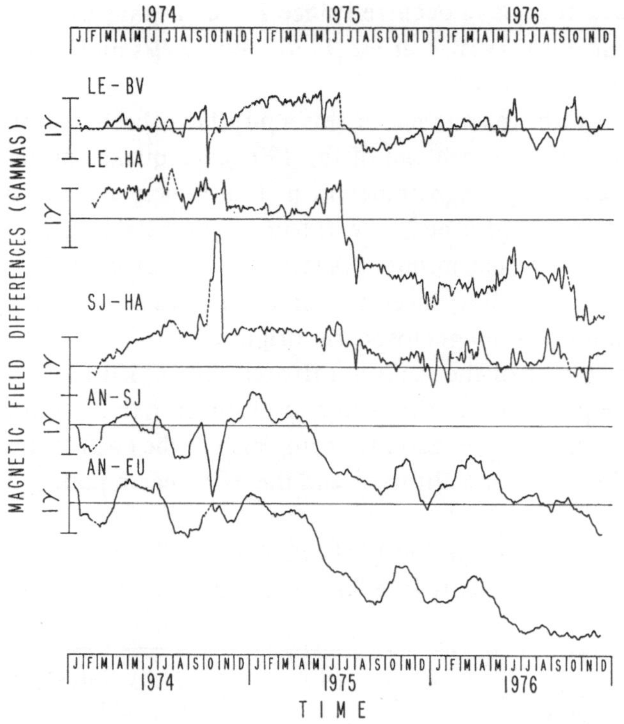

Fig. 3. Space-time history plots of magnetic field differences from station pairs to the southeast and the northwest along the fault from the point where aseismic slip ceases.

Fig. 3. A systematic decrease with time of the difference AN–EU and AN–SJ is clearly evident. The total change over the 3-year period, due primarily to a decreasing field at station AN, is about 2.5 γ. Independent data from repeated magnetic surveys between different sites near AN indicate a similar result (Johnston *et al.*, 1976).

There appear to be two possible explanations for these data. An obvious candidate is the secular variation of the geomagnetic field. In this case, however, direct effects of secular variation can be ruled out since (1) the apparent wavelength necessary would be a few tens of kilometers and (2) the change in the difference AN–SJ is opposite in sense to that expected for the present secular variation of about -30 γ/year. This arises since the net change in field at AN for large-wavelength, long-period geomagnetic field fluctuations is less than at other stations. The difference AN–SJ should increase for a 30-γ field decrease over 12 months in about the same way that it does during the 50-γ field decrease that occurs each day. For much the same reasons indirect or second-order effects of secular variation seem unlikely but cannot be ruled out.

A second possibility is a stress-induced magnetic change. This should be expected in this area if any substantial stress variation occurs. There are clearly insufficient data to attempt any rigorous inversion of the magnetic data to obtain the likely change in stress. The problem reduces therefore to finding the simplest model and inferred stress that will explain the data. This is treated in the next section.

3. Tectonomagnetic Model

Following CHINNERY (1963), PRESS (1965), and ROSENMAN and SINGH (1973), the stress distribution around a finite dislocation on a vertical fault can be calculated. These models relate the slip geometry and distribution to stresses and strains in the surrounding material. The details of the geometry and coordinate system are shown in Fig. 4. The fault has a length L. The top is at a depth d and the bottom is at a depth D. This type of model, and generalizations from it to variable slip and variable geometry (MCHUGH and JOHNSTON, 1978), have been used to calculate tilt, stresses, and strains for various slip geometries on the San Andreas fault.

In particular, models have been fit to (1) the general region near San Juan Bautista, where a transition from about 10% to less than 0.1% of the average plate displacement rate given in ATWATER and MOLNAR (1973) occurs in the surface fault-displacement measurements and (2) the general region on either side of the Cajon Pass end of the 1857 break where, historically, gradients in surface displacements have apparently occurred (SIEH, 1976). Only the simplest of these models, which can be applied to either of these two regions, is discussed here and is shown for the northern region in Fig. 5a.

In this model, the slip U is initially uniform and extends from San Juan Bautista to the southeast over the whole fault for at least 100 km. To the northwest, slip occurs only at depths below 11 km. Contours of maximum stress change (isopachs) derived from the sum of the principal stresses and normalized to $0.1U$, are plotted

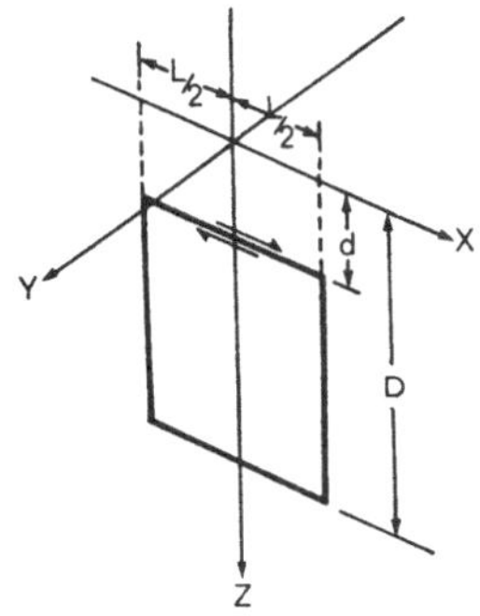

Fig. 4. Geometry and coordinate system used to model a vertical rectangular strike-slip fault.

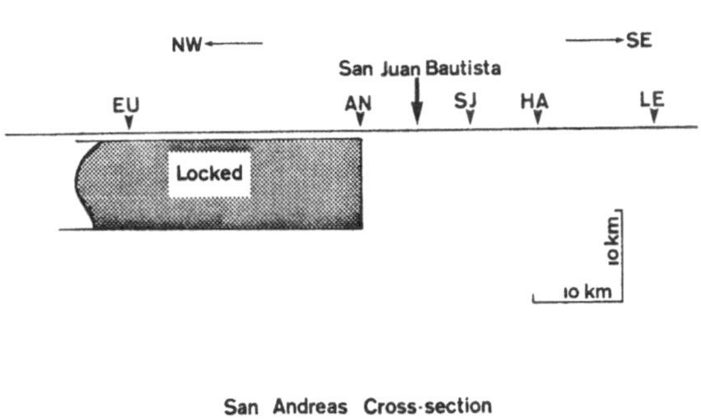

Fig. 5a. Simplified fault model near San Juan Bautista.

in Fig. 5b. The locations of recording magnetometers are shown as black dots. A slip of 100 cm gives shear-stress changes in the region around the end of as much as 20 bars. Quasistatic behavior can also be modeled but is not attempted here. For such a case, the slip boundary propagates along the fault together with the stress contour pattern, and a point on the surface would experience a transient stress field.

Following STACEY (1964), SHAMSI and STACEY (1969), and TALWANI and KOVACH (1972), the modification $\Delta \boldsymbol{I}$ of the total magnetization $\boldsymbol{I}(xyz)$ of a volume element of rock dv resulting from the imposition of a stress $\boldsymbol{\sigma}$ is found by first determining the components of magnetization in the direction of the principal stresses σ_1 from Eq. (1)

$$I_i=(D_{ji})\begin{pmatrix} I_x \\ I_y \\ I_z \end{pmatrix} \tag{1}$$

where D_{ij} is the direction cosine matrix with $j=x, y, z$ for $i=1, 2, 3$. Using the theoretically determined (STACEY and JOHNSTON, 1972) and experimentally supported (OHNAKA and KINOSHITA, 1968a, b) relation between change in magnetization ΔI_i in the σ_i direction, the values of ΔI_i can be found from:

$$\Delta I_i=\frac{C}{2}I_i(\sigma_j+\sigma_k-2\sigma_i) \quad i, j, k=1, 2, 3 \quad (i\neq j\neq k) \tag{2}$$

where C is the stress sensitivity ($\sim 10^{-4}\,\text{bar}^{-1}$; STACEY and JOHNSTON, 1972) and $i, j, k=1, 2, 3$, but $i\neq j\neq k$. The components ΔI_x, ΔI_y and ΔI_z in the x, y, z frame are then given by:

$$\begin{pmatrix} \Delta I_x \\ \Delta I_y \\ \Delta I_z \end{pmatrix}=(\Delta I_1, \Delta I_2, \Delta I_3)(D_{ji}) \tag{3}$$

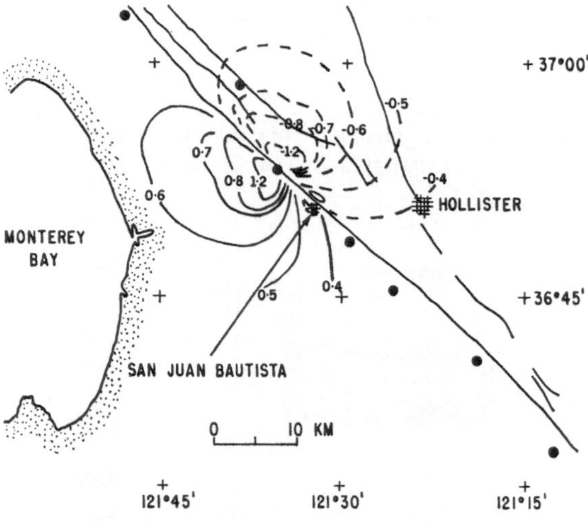

Fig. 5b. Contours of maximum stress ($\sigma_1+\sigma_2$) in bars at the northern end of the aseismically slipping section of the San Andreas fault normalized by $0.1U$ where U is the total fault slip in centimeters.

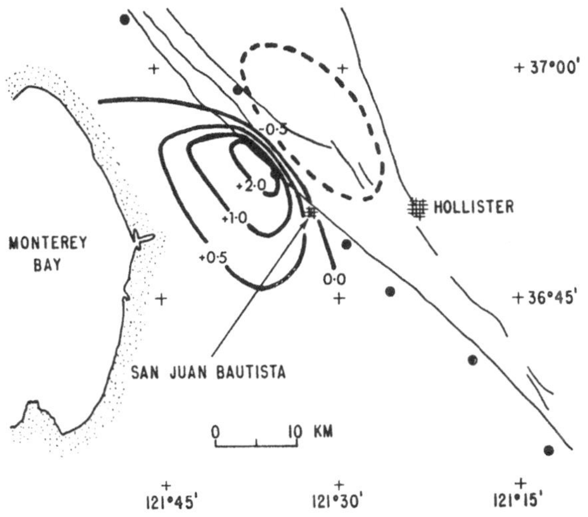

Fig. 5c. Contours of anomalous local magnetic field (γ) produced at the northern end of the slipping section of the San Andreas fault, if the region is assumed to have a magnetization of 10^{-3} emu.

and the surface magnetic anomaly components $\Delta F_{x,y,z}$ at a point $(x', y', 0)$ are determined (HILDERBRAND, 1975) by:

$$\Delta F_{x,y,z} = \int_v (\boldsymbol{\Delta I}\cdot\boldsymbol{\nabla}) \frac{d}{dx, y, z}\left(\frac{1}{r}\right) dv \tag{4}$$

where $r^2=(x-x')^2+(y-y')^2+z^2$. The surface total field anomaly can then be determined from these components.

The model in Fig. 5a can thus be used to determine the surface total field anomaly contours that presumably have been accumulating since at least the time of the 1906 earthquake. A slip of 3 cm/year for the last 70 years from a depth of 1 km to a depth of 11 km to the southeast of San Juan Bautista and below 11 km to the northwest produces the surface anomaly pattern shown in Fig. 5c. The contour interval is in 1-γ unit if the average magnetization is 10^{-3} emu and 10 γ if the average magnetization is 10^{-2} emu. The measured magnetizations of the surface rocks are typically about 10^{-3} emu but can exceed 10^{-2} emu in the installation region of station AN. If the remanent component is assumed to be oriented parallel to the induced component, the static magnetic anomaly indicated by the aeromagnetic map of the area (HANNA *et al.*, 1972) could be explained by stacked slabs of material with magnetizations generally increasing with depth from 10^{-4} emu to 10^{-3} emu except for a highly magnetized slab ($I_t \sim 10^{-3}$) about 5-km long, 3-km wide, and 5-km deep around station AN. The effects of introducing a more complex magnetization distribution complicates the surface anomaly pattern but not the general amplitudes indicated by the tectonomagnetic calculations.

4. Discussion

Determination of the origin of the observed 2-γ field decrease at station AN is difficult since tectonomagnetic models and attempts at inversion are poorly constrained with so few data over such a wide area. It is necessary to use other available data to limit the possibilities.

The observation of geodetically determined shear and volume strain in this region on lines 10 to 20 km in length (PRESCOTT and SAVAGE, 1978) independently limits the slip moment ($M=\mu AU$) allowed during the measurement period, where A is the area of slip and μ is the shear modulus. Since the average shear strain in the area from these data does not exceed 2 μstrain and the corresponding average change in shear stress on a 10-km scale is probably not more than 1 bar, the magnetic signal is almost certainly generated beneath and within a few kilometers of station AN. Until more magnetic as well as tilt and strain data are obtained in this area, the problem of interpretation of the magnetic data reduces to finding the simplest tectonomagnetic models that satisfy the data.

The simplest of these models involved uniform slip beneath the highly magnetized rock under AN. Two centimeters of slip over an area of about $0.5\times0.5\,km^2$ gives a release of shear stress of about 6 bars across the fault at a depth of 1 km beneath AN. Taking a uniform regional magnetization of 3×10^{-3} emu, a tectonomagnetic model with this geometry generates the required 2-γ negative anomaly at AN. Changes in strain and strain gradients within 500 m of the fault for this model would be as much as 8 and 1 μstrain in 100 m, respectively. Averaged over a 10-km geodetic line, however, the changes in strain are below the detection threshold of about 0.5 μstrain for the geodetic measurements (PRESCOTT and SAVAGE, 1978).

Encroachment of slip into the locked segment of the fault implied by this model might be expected on the basis of plots of cross sections of the seismicity of this area taken from WESSON *et al.* (1973) and shown in Fig. 6. The microearthquake distribution is most intense in the region where the fault is slipping but falls off and apparently dips under the region near AN where the fault is locked. The slip patch indicated by this model is shown as a square beneath AN.

Other models in which the slip area is larger or where the slipping region is deeper or some combination of these variables are, of course, possible. In general, increasing the depth for fixed geometry requires rapidly escalating stress amplitudes to give the same surface anomaly, and increasing the area of the slip patch by more than a factor of 3, implies geodetically detectable strains. The important implication of these models therefore remains unchanged. This implication is that, if the magnetic data are to be explained by a stress perturbation, the source is probably quite localized (i.e. less than 5 km from AN), and the change in stress in not more than about 10 bars.

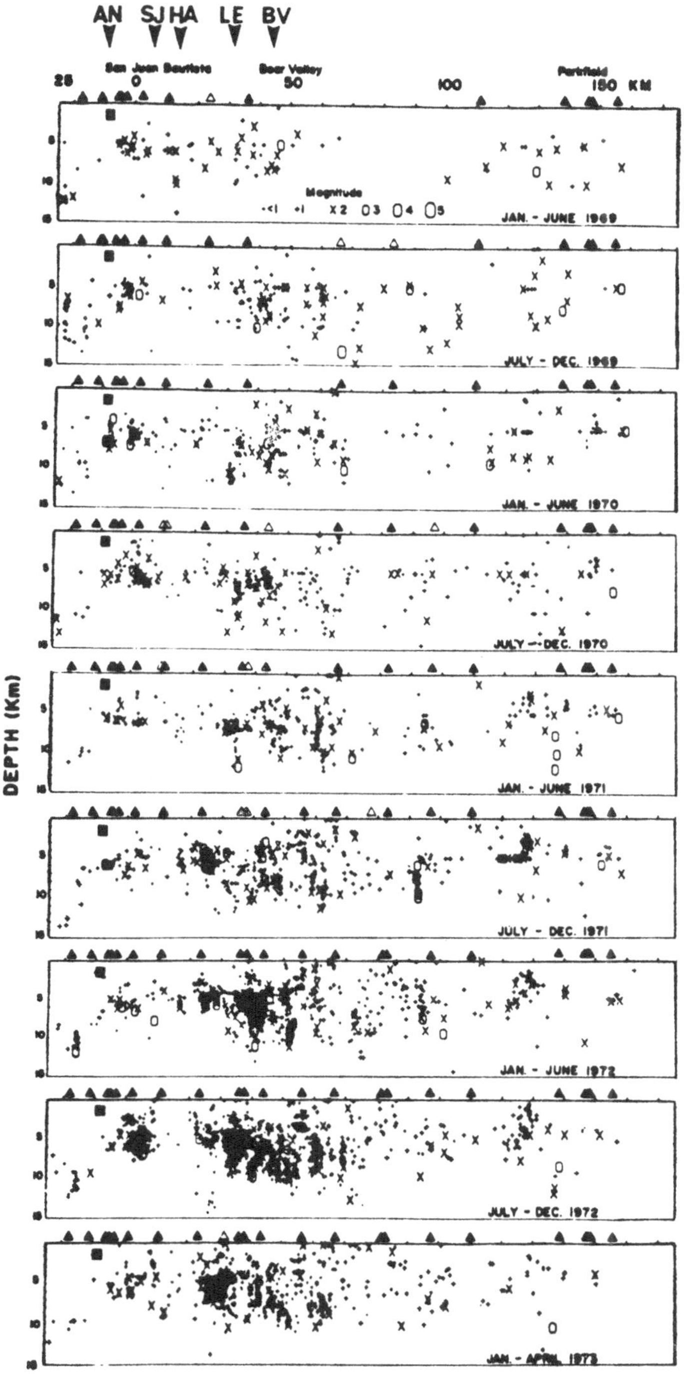

Fig. 6. Cross sections of the seismicity along the San Andreas fault during consecutive 6-month intervals from 1969 to April 1973. Solid triangles denote seismograph stations within 10 km of the fault in operation during the entire period covered by the section, and open triangles denote stations installed or removed during the period (after WESSON *et al.*, 1973). The magnetometer locations are marked with arrows and the location of the slip patches proposed to explain the data is shown in squares on the cross section.

5. Conclusions

The systematic decrease in local magnetic field at present occurring near San Juan Bautista on the San Andreas fault constitutes one of the most significant well-recorded field perturbations yet obtained along the fault. It is in this region that the 1906 rupture stopped, and a transition occurs from a stable sliding fault to the southeast to a locked and generally aseismic fault to the northwest.

Determination of the physical origin of the observed change is hindered by the paucity of data. An accumulated field change of about 2γ would be expected throughout this general region if the presently observed rate of strain accumulation is assumed to have been uniform for the past 70 years since the 1906 earthquake. However, the observed change is occurring too rapidly to be due to the average accumulation of strain, and more significantly, has an opposite sign. Furthermore, the absence of any abnormal large-scale strains in the geodetic data in this area (PRESCOTT and SAVAGE, 1978) limits the amount of aseismic slip during the observational period. The simplest class of models that explain the data invokes localized and quite near-surface systematic encroachment of slip into the locked section of the fault. Other possibilities such as a growing but localized stress concentration or heterogeneity could be suggested but cannot presently be independently supported. Regardless of whether fault slip or a local stress concentration is occurring, at least 5-bars change in stress is required to explain the magnetic data if the source is within a few kilometers of the earth's surface.

It is fortunate, both for determining the reality of a mechanical origin for the magnetic observations and for quantifying the models suggested, that a number of new magnetometers and other types of instruments (strain, tilt) are now installed in the region just north of San Juan Bautista. Should a moderate magnitude earthquake occur, for the first time, changes in stress estimated by different techniques can be compared.

I thank Bruce Smith for providing, unpublished data, Dr. Leroy Allredge for an unpublished preprint, and Allan Lindh for an unpublished integration of California seismic data and stimulating discussion.

REFERENCES

ATWATER, T. and P. MOLNAR, Relative motion on the Pacific and North American plates deduced from sea floor spreading in the Atlantic, Indian and South Pacific oceans, in *Proceedings of the Conference on Tectonic Problems of the San Andreas Fault System*, edited by R.L. Kovach and A. Nur, pp. 136–148, School of Earth Sciences, Stanford University, Stanford, California, 1973.

BOLT, B.A. and R.D. MILLER, Catalogue of earthquakes in northern California and adjoining Areas—January 1910–13 December 1972, Bulletin of the Seismographic Stations, University of California, Berkeley, 568 pp., 1975.

BRUNE, J.N., Tectonic stress and the spectra of seismic shear waves from earthquakes, *J. Geophys. Res.*, **75**, 4997–5009, 1970.

BRUNE, J.N., T.L. HENYEY, and R.F. ROY, Heat for, stress, and rate of slip along the San Andreas fault, California, *J. Geophys. Res.*, **74**, 3821–3828, 1969.

CHINNERY, M.A., The stress changes that accompany strike-slip faulting, *Bull. Seismol. Soc. Am.*, **53**, 921-932, 1963.

CHINNERY, M.A. and J.A. PETRAK, The dislocation model with variable discontinuity, *Tectonophysics*, **5**, 513-529, 1968.

FRIEDMAN, M.E., J.H. WHITCOMB, C.R. ALLEN, and J.A. HILEMAN, Seismicity of the Southern California Region, 1 January 1972 to 31 December 1974, Seismological Laboratory, California Institute of Technology, Pasadena, California, 81 pp., 1976.

HANNA, W.F., R.D. BROWN, D.C. ROSS, and A. GRISCOM, Aeromagnetic reconnaissance along the San Andreas fault between San Francisco and San Bernardino, California: U.S. Geological Survey, Geophys. Inv. Map GP 815, scale 1:250,000, 1972.

HILDERBRAND, T., Seismomagnetism, Ph.D. thesis, University of California, Berkeley, 1975.

HILEMAN, J.A., C.R. ALLEN, and J.M. NORDQUIST, Seismicity of the southern California region 1 January 1932 to 31 December 1972, Seismological Laboratory, California Institute of Technology, Pasadena, 404 pp., 1973.

JOHNSTON, M.J.S., B.E. SMITH, and R. MUELLER, Tectonic experiments and observations in western U.S.A., *J. Geomag. Geoelectr.*, **28**, 85-97, 1976.

LACHENBRUCH, A.H. and J.H. SASS, Thermo-mechanical aspects of the San Andreas fault, in *Proceedings of the Conference on Tectonic Problems of the San Andreas Fault System*, edited by R.L. Kovach and A. Nur, pp. 192-205, School of Earth Sciences, Stanford University, Stanford, California, 1973.

MCHUGH, S. and M.J.S. JOHNSTON, Dislocation modelling of creep related tilt changes, *Bull. Seismol. Soc. Am.*, **68**, 155-168, 1978.

OHNAKA, M. and H. KINOSHITA, Effect of uniaxial compression on Remanent Magnetization, *J. Geomag. Geoelectr.*, **20**, 93, 1968a.

OHNAKA, M. and H. KINOSHITA, Effect of axial stress upon initial susceptibility of an assemblage of fine grains of Fe_2TiO_4-Fe_3O_4 solid solution series, *J. Geomag. Geoelectr.*, **20**, 107, 1968b.

PRESCOTT, W. and J.C. SAVAGE, Strain accumulation rates in the western United States between 1970 and 1976, *J. Geophys. Res.*, 1978 (in press).

PRESS, F., Displacements, strains and tilts at teleseismic distances, *J. Geophys. Res.*, **70**, 2395-2412, 1965.

ROSENMAN, M. and S.J. SINGH, Quasi-static strains and tilts due to faulting in a viscoelastic halfspace, *Bull. Seismol. Soc. Am.*, **63**, 1737-1742, 1973.

SHAMSI, S. and F.D. STACEY, Dislocation models and seismomagnetic calculations for California 1906 and Alaska 1964 earthquakes, *Bull. Seismol. Soc. Am.*, **59**, 1435-1448, 1969.

SIEH, K., Pre-historic large earthquakes on the San Andreas fault in the area of the Palmdale Uplift, *EOS, Trans. Am. Geophys. Union*, **57**, 899, 1976.

SMITH, B.E. and M.J.S. JOHNSTON, A tectonomagnetic effect observed before a magnitude 5.2 earthquake near Hollister, California, *J. Geophys. Res.*, **81**, 3556-3560, 1976.

SPOTTISWOOD, S.M. and A. MCGARR, Source parameter of tremors in a deep-level gold mine, *Bull. Seismol. Soc. Am.*, **65**, 93-112, 1975.

STACEY, F.D., The seismomagnetic effect, *Pure Appl. Geophys.*, **58**, 5-22, 1964.

STACEY, F.D. and M.J.S. JOHNSTON, Theory of the piezomagnetic effect in titanomagnetite bearing rocks, *Pure Appl. Geophys.*, **97**, 146-155, 1972.

STESKY, R.M. and W.F. BRACE, Estimation of frictional stress on the San Andreas fault from laboratory measurements, in *Proceedings of the Conference on Tectonic Problems of the San Andreas Fault System*, edited by R.L. Kovach and A. Nur, pp. 206-213, Stanford University Publications, 1973.

TALWANI, P. and R.L. KOVACH, Geomagnetic observations and fault creep in California, *Tectonophysics*, **14**, 245-256, 1972.

TOPPAZADA, T.R., D.L. PARKE, and C.T. HIGGINS, Seismicity of California, 1900-1931, unpublished manuscript, 1976.

WESSON, R.L., R.O. BURFORD, and W.L. ELLSWORTH, Relationship between seismicity, fault creep and crustal loading along the central San Andreas fault, in *Proceedings of the Conference on Tectonic*

Problems of the San Andreas Fault System, edited by R.L. Kovach and A. Nur, pp. 303–321, School of Earth Sciences, Stanford University, Stanford, California, 1973.

WILLIAMS, F.J. and M.J.S. JOHNSTON, Differential magnetometer measurements on the Palmdale Bulge, and along the San Jacinto fault, *EOS, Trans. Am. Geophys. Union*, **57**, 898, 1976.

Geomagnetic Secular Variation Anomalies in the GDR

Wolfgang MUNDT

Central Institute of Earth Physics, 15 Potsdam, Telegrafenberg, GDR

(Received December 6, 1977)

For about 25 years, measurements of the magnetic components have been carried out at repeat stations in the territory of the GDR. All measurements at common stations of the magnetic surveys for the epochs 1901, 1935, and 1957 were compared. The results of the investigation indicate secular variation anomalies with amplitudes of about 3–4 nT/year. These anomalies are obviously partly located in regions with marked recent vertical and horizontal movements. Moreover they seem to be characterized by positive anomalies of the heat flow.

1. Introduction

The detection of regional and local anomalies of the geomagnetic secular variation presents problems in regions with large station separations and low observation accuracies in relation to the small dimensions and low amplitudes of these anomalies.

This problem is also reflected by the investigations of the distribution of SV anomalies in the GDR (BOLZ and KAUTZLEBEN, 1962; MUNDT, 1974a). A joint evaluation of all measurements available has provided new information about the secular variation anomalies in the territory of the GDR.

2. Observational Data

For the territory of the GDR, three magnetic land surveys are available for the epochs 1901 (SCHMIDT, 1910), 1935 (BOCK *et al.*, 1948), and 1957 (BOLZ *et al.*, 1969).

Moreover the secular variation of the magnetic field is observed by means of a network of 11 repeat stations and at the Niemegk observatory. To supplement these data, the declination D and the horizontal component H of about 50 stations for the 1957 epoch were remeasured for the 1976 epoch (WOLTER, 1976). Figure 1 shows the distribution of the repeat stations and the common points for the 1957 and 1976 epochs.

The behaviour of the secular variation at the repeat stations is shown by Figs. 2, 3, and 4. The similarity of the profiles indicates a nearly uniform change of the secular variation in the territory of the GDR. Anomalous areas are defined by the very small differences in the slopes of the curves at the repeat stations, for example at Groß Werther, Schleusingen, and Jöhstadt. They cannot, however, be described

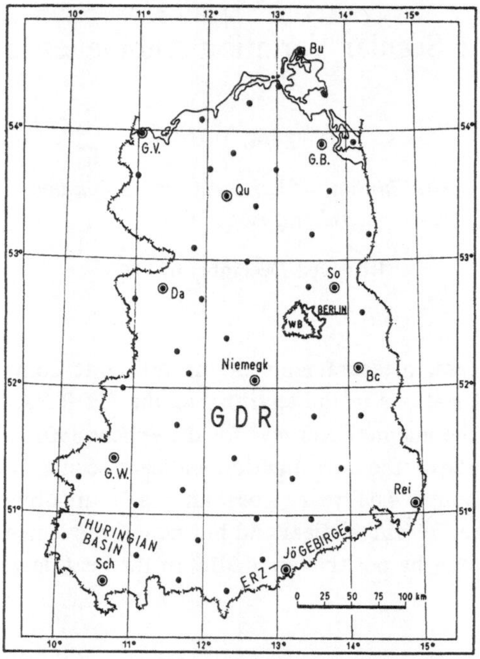

Fig. 1. Distribution of repeat stations in the GDR.

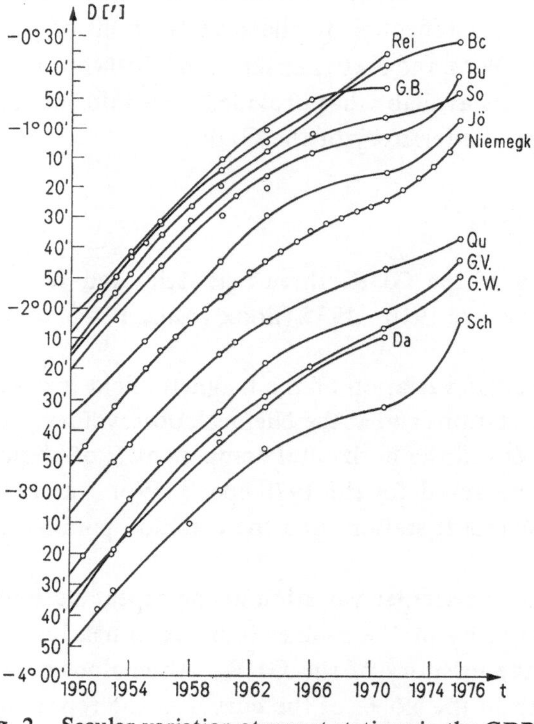

Fig. 2. Secular variation at repeat stations in the GDR Declination.

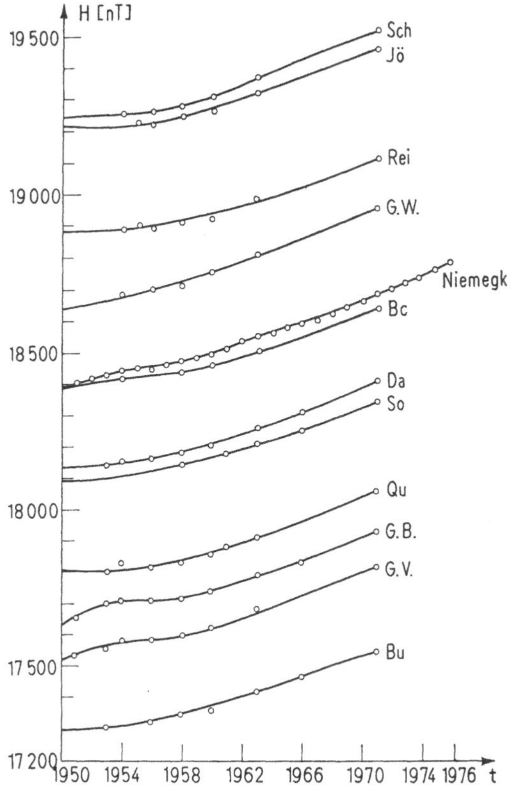

Fig. 3. Secular variation at repeat stations in the GDR *H*-component.

exactly on the basis of the repeat stations alone. It should be emphasized that the secular variation observed at the Niemegk observatory provides a good approximation to the average behaviour in the territory of the GDR.

3. Anomalies of Secular Variation for *D* and *H*

The regional structure of the SV in the declination and in the horizontal intensity was determined from the measurements referred to the 1957 and 1976 epochs for the stations shown in Fig. 1. Here the mean station distance is about 50 km. Moreover the data obtained for the 1901 and 1935 epochs were used when the accuracy requirements were satisfied.

Figures 5 and 6 show the absolute variations of *D* and *H* for the time interval of 19 years. The shown isopores are the result of a generalization according to the method of error margins, which was based on mean errors of $\pm 2'$ for *D* and ± 10 nT for *H*. Not shown are all those anomalies whose amplitudes lie within these error intervals.

Three anomalous areas appear on the *D*-map, while on the *H*-map two anomalies occur in the Thuringian basin and in the western Erzgebirge.

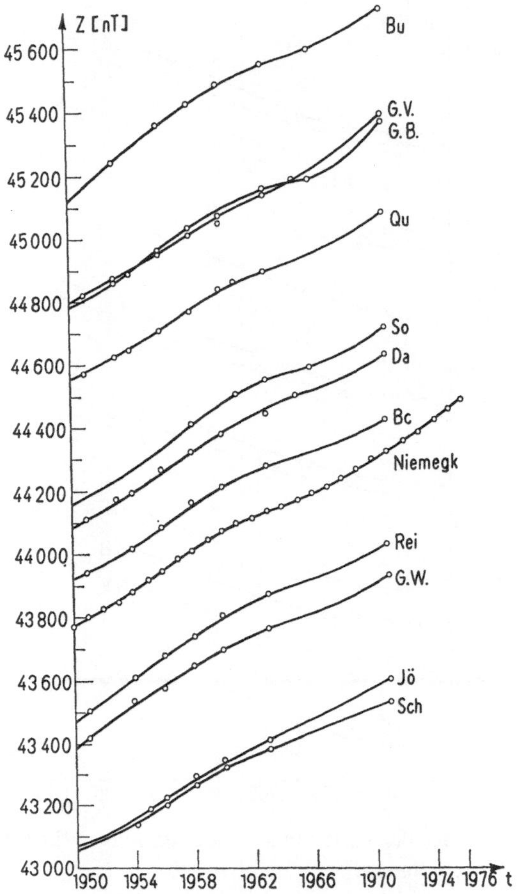

Fig. 4. Secular variation at repeat stations in the GDR *Z*-component.

4. Anomalies of Secular Variation for *F*

In addition to the repeat stations, we used all stations where magnetic data were available for at least two of the epochs 1901, 1935, and 1957. The necessary conversion of the data into the 'total field variation per year' was carried out subject to the assumption that the variation of the total intensity on the GDR territory was almost linear in the time intervals 1901–1925 and 1925–1976, as this was the case with the secular variation observed at the Niemegk observatory. A total of 80 data points were available for describing the regional structure of secular variation in *F*. This corresponds to a station distance of about 40 km.

Since the definition of the optimum secular variation trend is not unique, the anomalies are calculated as referred to two different trends. The first trend was a 1st order polynomial calculated by curve fitting of the SV data for the territory of the GDR according to the least squares method. The second trend was a 3rd order polynomial calculated by a corresponding curve fitting of the SV data obtained at

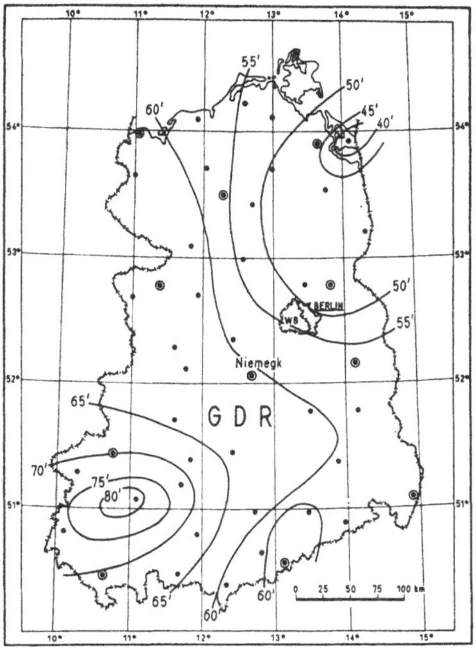

Fig. 5. Secular variation in declination, 1957–1976.
Units: minutes.

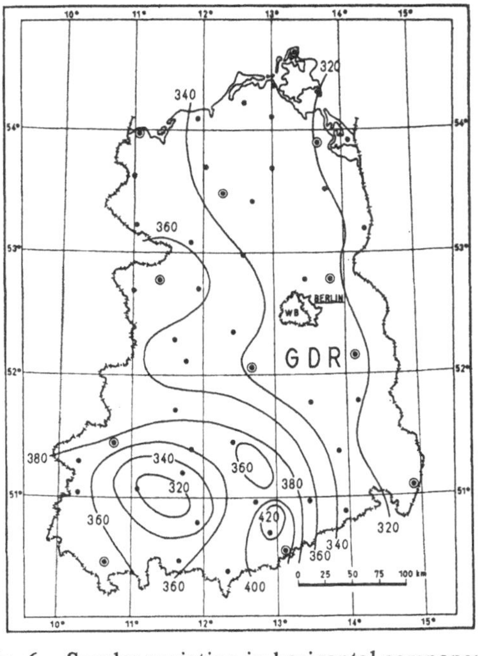

Fig. 6. Secular variation in horizontal component,
1957–1976. Units: nT.

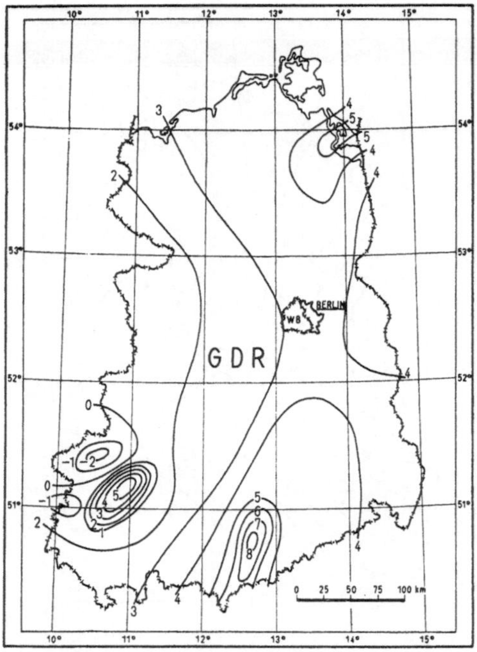

Fig. 7. Secular variation anomalies $\delta \Delta F$ (nT/year). Trend: 1° degree polynomial (repeat stations).

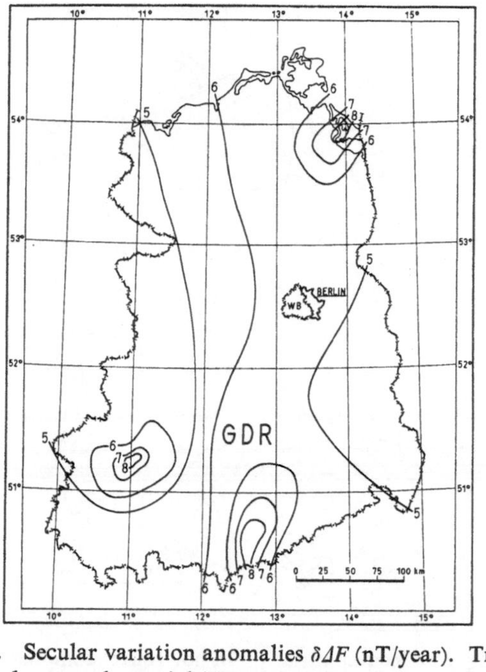

Fig. 8. Secular variation anomalies $\delta \Delta F$ (nT/year). Trend: 3° degree polynomial (European observatories).

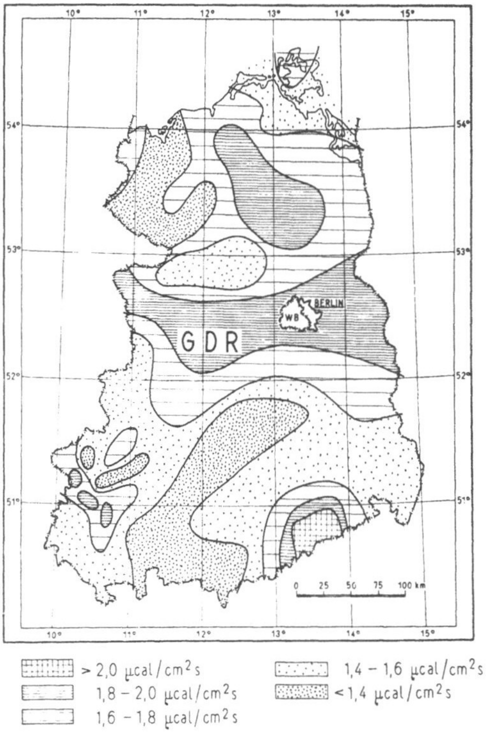

Fig. 9. Regional pattern of the heat flow (after Hurtig).

the European observatories. Figures 7 and 8 show the SV anomalies for F. The isopores are the result of an objective smoothing according to the method of error margins, which was based on a mean error of ± 0.5 nT/year. Anomalies which had amplitudes of <1 nT/year were not taken into account.

In the two maps three local anomalies appear, which were indicated in the maps for D and H. The mean dimension of the anomalies amount to 50–100 km.

5. Anomalies of Secular Variation and Heat Flow

A comparison of the anomalies of the secular variation with the map of heat flow according to Hurtig (Hurtig and Schlosser, 1975) shows striking correspondence with respect to the location of the anomalies in the south of the GDR (Fig. 9). It is observed that obviously positive anomalies of the secular variation for F are correlated with positive anomalies of the heat flow, even if all of the possible inaccuracies are taken into account.

6. Anomalies of Secular Variation and Recent Movements of the Earth' Crust

It is of interest to look for relationships between SV anomalies and recent movements of the earth's crust. In this respect there are favourable conditions for the territory of the GDR. It is possible to make use of the maps of recent vertical crust movements and their gradients, which are available for the GDR territory (THURM *et al.*, 1965; THURM, 1971). Moreover detailed investigations were performed of the horizontal dislocations and deformations of the earth's crust for several regions in the south of the GDR, where the strain parameters were calculated in order to characterize strains, compressions, and shearings (THURM, 1974, 1977).

From these data the following picture resuts:

The anomaly of secular variation occurring in the southwest of the GDR lies in a region of relatively rapid subsidence (−2.5 to −3.5 mm/year), while in the region of the Erzgebirge anomaly subsidence occurs as well, but with different amounts ranging from −1.5 to −2.5 mm/year. The maps of the gradients of the recent vertical crust movements and the tectonic faults in Europe (JUBITZ, 1976) indicate that especially the region of the west Erzgebirge is characterized by a great number of gradients and geologic faults of different extension and orientations.

The horizontal crust movements have amplitudes of about 5 to 10 mm/year. The degree of differentiation of these movements is indicated by the distributions of the main strain and the rotation component which are shown in Fig. 10. Extension is found to occur throughout the core of the anomaly, whereas compression zones are encountered at the northwest and east boundaries. Moveover, rotational movements occur in the region of the anomaly.

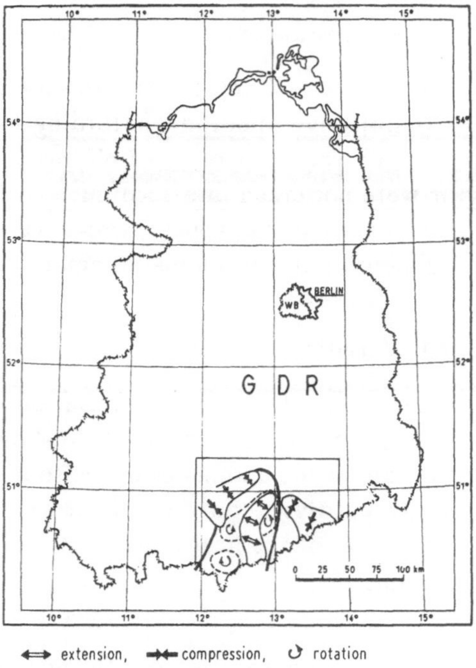

Fig. 10. Main strain and component of rotation in the area of secular variation anomaly (after Thurm).

7. Interpretation

The anomalies of secular variation in the south of the GDR are obviously partly located in regions with marked recent vertical and horizontal movements. Moreover they seem to be characterized by positive anomalies of the heat flow. However, it is too early to draw any definite conclusions. An interpretation of the physical processes within the anomalous regions is still lacking also. Due to the indicated considerable horizontal shifts and rotations it is likely that the positive SV anomalies in the Erzgebirge cannot be explained only by a decrease in the depth of the Curie isotherm which is associated with the subsidence.

Obviously the situation is much more complicated, and certainly magneto-mechanic phenomena will also play a role. However, for establishing a quantitative model it will be necessary to perform a survey of the fine structure of the anomalies using high-precision proton-vector magnetometers over a period of several years.

REFERENCES

Bolz, H. and H. Kautzleben, Anomalien der geomagnetischen Säkularvariation von 1935, 0 bis 1957, 5 auf dem Gebiet der DDR, Jahrbuch, Obs. Niemegk, 121–218, Berlin, 1962.

Bolz, H., H. Kautzleben, W. Mundt, and H. Wolter, Die magnetische Landesvermessung der Deutschen Demokratischen Republik zur Epoche 1957, 5; Ergebnisse und Auswertung, *Abh. Geomagnet. Inst. Potsdam*, Nr. 41, 197–219, 1969.

Bock, R., F. Burmeister, and F. Errulat, Magnetische Reichsvermessung 1935, 0, *Abh. Geophys. Inst. Potsdam*, Nr. 6, 1–53, 1948.

Hurtig, E. and P. Schlosser, Untersuchung des terrestrischen Wärmeflusses in der DDR, *Gerl. Beitr. Geophys.*, **84**, 235–246, 1975.

Jubitz, K.-B., Tektonische Bruchstörungen in Europa, Karte M. 1: 6 Mill., *Veröff. Zipe*, Nr. 47, Karte 7, 1976.

Mundt, W., Säkularreduktion der geomagnetischen Landesaufnahme der DDR zur Epoche 1957, 5, *Gerl. Beitr. Geophys.*, **83**, 147–158, 1974a.

Mundt, W., Zur Signifikanz und geologischen Deutbarkeit magnetischer Säkularvariationsanomalien in Europa, *Veröff. Zipe*, Nr. 23, 1–23, 1974b.

Schmidt, A., Magnetische Karten von Norddeutschland für 1909, *Veröff. Phys. Met. Inst.*, Abstr. III, 4, 1910.

Thurm, H., H. Moutag, A. Lang, P. Baukwitz, L. Hievsemann, W. Spouheuer, and E. Neef, Vorläufige karte der rezenten vertikalen Krustenbewegungen in der Deutschen Republik, *Peterm. Geogr. Mitt.*, H. 2, 136–160, 1965.

Thurm, H., Ableitung von Gradienten rezenter vertikaler Erdkrustenbewegungen für des Gebiet der DDR, *Peterm. Geogr. Mitt.*, H. 2, 124–129, 1971.

Thurm, H., Horizontale Dislokationen und Deformationen in der Elbtalzone, *NKGG-Veröff.*, Reihe III, H. 35, 6–20, 1974.

Thurm, H., Die aus älteren Triangulationen abgeleiteten horizontalen Deformationen und Spannungen der Erdkruste im Südostteil der Deutschen Demokratischen Republik, *Peterm. Geogr. Mitt.*, H. 4, 281–288, 1977.

Wolter, H., Ergebnisse der Vermessung von *D* und *H* an 50 Säkularpunkten im Gebiet der DDR, Unveröffentlichter Bericht, 1976.

Noise Reduction Techniques for Use in Determining Local Geomagnetic Field Changes

R.H. WARE and P.L. BENDER*

Joint Institute for Laboratory Astrophysics, University of Colorado and National Bureau of Standards, Boulder, Colorado 80309, U.S.A.

(Received October 10, 1977)

Measurements of the difference in total field $\Delta F(t)$ have been made over a 16 km N-S path near Boulder. The behavior observed is quite different from that for an E-W path. The present accuracy of the narrow line rubidium magnetometers used is about 0.01 γ. The N-S variations appear to correlate mainly with variations in H rather than D, and may be associated with either gradients of external fields or currents in shallow conductivity anomalies. More recently three magnetometers have been set up on a straight E-W line so that the 'second difference' can be measured, and a transfer function from field component variation to the second difference can be determined. A generalization of this approach will be used for analyzing USGS tectonomagnetic data from California.

1. Introduction

Recent years have seen a burgeoning interest in earthquake prediction. In view of the fact that a redistribution of crustal stress may precede most earthquakes, the ability to measure changing stress is an important prognostic tool. Efforts to use magnetometers as remote sensors of crustal stress changes have had some success in the past few years (SMITH and JOHNSTON, 1976). However, magnetic manifestations of changing stress can be obscured by a background of magnetic signals due to geophysical events not related to earthquakes (JOHNSTON *et al.*, 1976). Efforts to understand and reduce this background using data from highly accurate (0.01 γ) rubidium narrow line magnetometers are discussed below.

2. Instrumentation

Narrow line Rb^{87} magnetometers (ALLEN and BENDER, 1972; BRILL, 1975; BEAHN, 1976), which will be referred to as NLMs, have been used to obtain geomagnetic data near Boulder, Colorado. An improvement to previously reported versions of the instrument was accomplished recently by replacing the aluminum can which houses the NLM with a lucite can. This modification eliminated a troublesome temperature-dependent diurnal drift of about 0.05 γ. The cause of the drift is

* Staff Member, Quantum Physics Division, National Bureau of Standards.

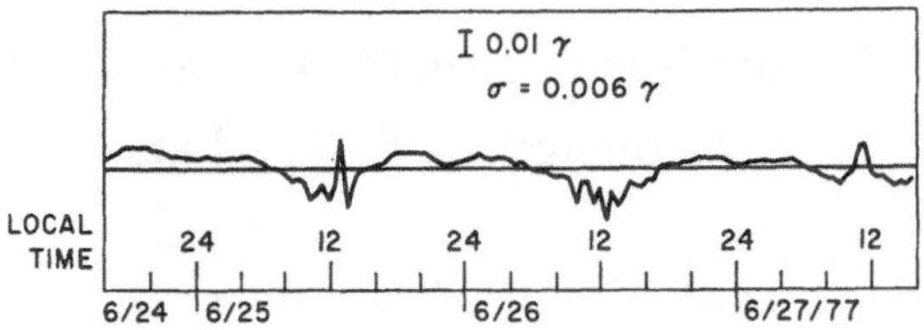

Fig. 1. Three days of total field differences measured by two NLMs separated by 20 cm averaged over half-hour intervals. The standard deviation of the half-hour averages is 0.006 γ.

poorly understood, but it appeared to be due to thermoelectric currents flowing in the aluminum.

Three days of observations using two of the improved instruments separated by 20 cm are shown in Fig. 1. The standard deviation of the half-hour averages of total field difference is 0.006 γ. The accuracy is now believed to be 0.01 γ rms in terms of the Rb^{87} Zeeman frequency. An additional uncertainty of 0.007 γ is involved in converting to the IUGG 1960 magnetic field scale (ALLEN and BENDER, 1972). However, this does not affect ΔF, and hence is unimportant for the observation of anomalies.

3. Gradiometer Measurements

Initial field tests of NLMs using instruments at each end of a 12 km east–west (EW) baseline revealed strong coherence between the declination component D of the field and the difference in total field between the two stations, ΔF, for periods ranging from 0.5 to 24 hr (BRILL, 1975). Data taken consisted of intervals ranging from 1 to 49 hr during 11 days of measurements. D, which was measured by the Boulder Observatory, is roughly perpendicular to the high conductivity structure under the Rocky Mountains (GOUGH, 1974) and to the western edge of the Denver Basin sediments. Brill suggested that the coherence between D and ΔF(EW) can be attributed to differences in the fields at the two sites due to currents induced in the north–south conductivity structures by fluctuations in D.

Spectral analysis of subsequent measurements of ΔF(EW) for 41 consecutive days led to the calculation of a frequency-dependent transfer function from D to ΔF(EW) by BEAHN (1976). The horizontal and vertical components H and Z were found to have poor coherence to ΔF(EW). An induced ΔF was calculated using D and the transfer function, and was subtracted from the observed ΔF. This gave a reduction in the average standard deviation of daily averages for 41 days from 0.21 to 0.08 γ.

A third instrument has been set up to measure ΔF over a 16 km north–south (NS) baseline. Data are relayed by telemetry and recorded on magnetic tape every 10 sec. All instruments are the improved variety mounted in lucite cans described above. Other details of the experimental setup are available elsewhere (BEAHN, 1976). Typical behavior for 24 quiet hours is shown in Fig. 2. Preliminary analysis

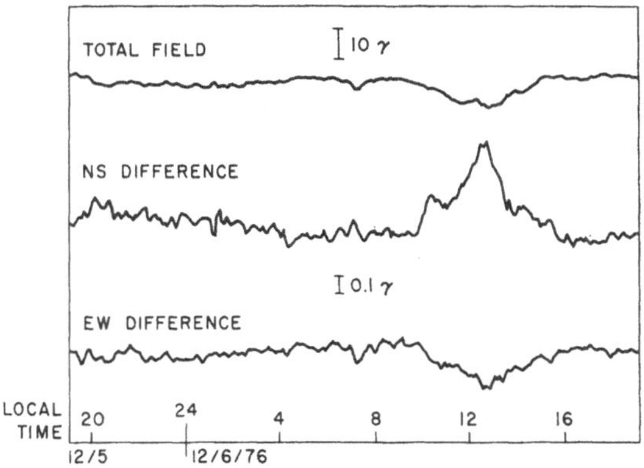

Fig. 2. Six-minute averages of total field, 16 km north–south difference field, and 12-km east–west difference field for 24 hr during quiet times.

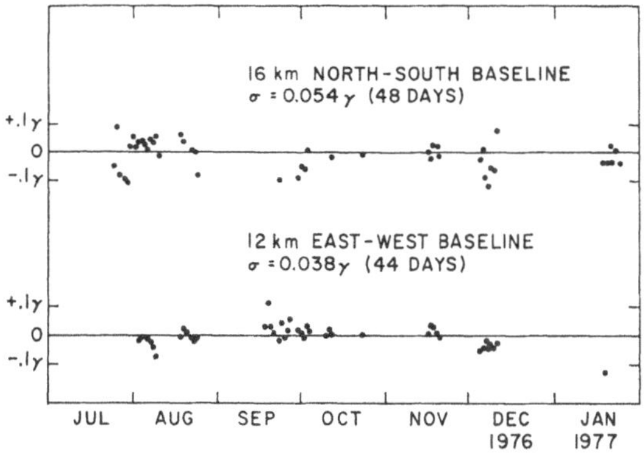

Fig. 3. Daily averages of total field difference for measurements taken during a 6-month period.

of 7 days of continuous data reveals strong coherence (~ 0.8) from H to ΔF(NS) for periods of 0.5 to 3.5 hr and poor coherence for other periods. The components D and Z show poor coherence to ΔF(NS). Daily averages of data ranging over a 6-month period are shown in Fig. 3.

In general, frequency-dependent coherence between field components and ΔF is not caused by differences in susceptibility or total magnetization directions between sites, as magnetic signals attributed to these sources should result in coherence which is independent of frequency. Therefore, considering other possible sources of magnetic signals (Johnston *et al.*, 1976), the high coherence between H and ΔF(NS) which was found only for higher frequencies suggests external fields or externally induced internal currents in relatively shallow conductivity anomalies as signal sources. From the standpoint of noise reduction, the periods ranging from 0.5 to 3.5 hr which show high coherence from H to ΔF(NS) contain a large fraction of the total power, and the transfer function approach is therefore expected to have some success in reducing noise. However, this approach by itself does not seem capable of strongly reducing the noise at longer periods, which are of the main interest for earthquake prediction research. The transfer function treatment for the NS baseline has not yet been completed, and further discussion must be left to future publications.

4. Second Difference Measurements

Comparison of magnetograms from Boulder and Tucson during unsettled conditions shows field gradients giving $>0.1\ \gamma$ field difference for a 16 km NS baseline. Thus the field gradients in the NS direction due to the ionospheric current geometry may be significant a substantial part of the time. This could explain the observed low coherence at long periods between ΔF(NS) and the variations in the components of the local field. In the absence of a deep conductivity anomaly having the right geometry, there is no clear reason for variations in the gradient to be correlated with variations in the components.

In an effort to separate the effect of these large scale gradients from local effects, three instruments were set up with 12 km separation on a straight EW line. Consider F_1, F_2, F_3 to be the scalar total field measured at the western, center, and eastern stations. The second difference is then defined as $F_2-1/2(F_1+F_3)$. The results of a second difference calculation for 6-min averages of 24 hr of data taken during active times are shown in Fig. 4. Two days of data have been analyzed so far, and the same striking noise reduction and the suggestion of moderately long period structure occurring near 13:00 local time is apparent in both. We are curious about the cause of this structure and hope to learn more about it by using larger blocks of second difference data and applying the transfer function technique to the second difference residuals. We expect that by combining the transfer function and second difference approaches in this way, we can further reduce the variations in the results.

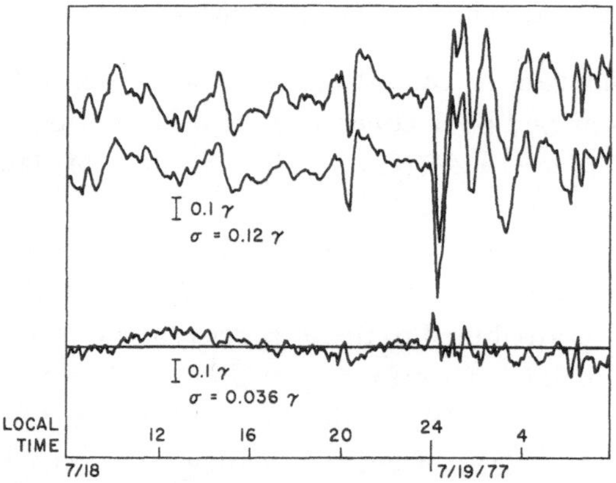

Fig. 4. Six-minute averages of data recorded during quiet to active times from three stations separated by 12-km intervals on a straight EW line. Starting at the top, the curves are F_1-F_2 and $1/2(F_1-F_3)$. The lower curve is the difference of the upper curves, $F_2-1/2(F_1+F_3)$, which we define as the second difference.

5. Conclusions

The analysis of data taken on the NS baseline described above is currently incomplete, as is the analysis of the second difference data, and conclusions are not yet warranted. However, the results of BEAHN (1976) and the preliminary results presented above have demonstrated the ability of the transfer function and of the second difference technique to reduce noise for data taken in Colorado to a level of 0.04 to 0.08 γ over our baselines. We hope that the combined transfer function/second difference approach will decrease the variations in the daily averages to about the 0.01 γ accuracy of our instruments. Efforts to apply an extension of these techniques to data taken by the USGS near the San Andreas Fault in central California are being started.

REFERENCES

ALLEN, J.H. and P.L. BENDER, Narrow line rubidium magnetometer for high accuracy field measurements, *J. Geomag. Geoelectr.*, **24**, 105–125, 1972.

BEAHN, T.J., Geomagnetic field gradient measurements and noise reduction techniques in Colorado, *J. Geophys. Res.*, **81**, 6276–6280, 1976.

BRILL, R., Differential geomagnetic field measurements at the edge of the Denver Basin, *J. Geophys. Res.*, **80**, 1593–1599, 1975.

GOUGH, D.I., Effects of induction in the earth upon measurements of external time-varying magnetic fields, *EOS, Trans. Am. Geophys. Union*, **55**, 595–599, 1974.

JOHNSON, M.J.S., B.E. SMITH, and R. MUELLER, Tectonomagnetic experiments and observations in Western U.S.A., *J. Geomag. Geoelectr.*, **28**, 85–97, 1976.

SMITH, B.E. and M.J.S. JOHNSTON, A tectonomagnetic effect observed before a magnitude 5.2 earthquake near Hollister, California, *J. Geophys. Res.*, **81**, 3556–356, 1976.

Local Variations in Magnetic Field, Long-Term Changes in Creep Rate, and Local Earthquakes along the San Andreas Fault in Central California

B.E. SMITH, M.J.S. JOHNSTON, and R.O. BURFORD

U.S. Geological Survey, 345 Middlefield Road, Menlo Park, California 94025, U.S.A.

(Received December 1, 1977)

Comparison between local variations in magnetic field, long-term changes in creep rate, and local earthquakes have been made for the seismically active and creeping section of the San Andreas fault between the most southern extent of the 1906 earthquake fault break and the most northern extent of the 1857 break, for the period early 1974 through mid-1977. The data utilized are from stations located near the two ends of this section of the San Andreas fault where strain accumulation is expected. The proton precession magnetometer stations included in this study have recorded local magnetic field variations up to 1.8 γ with durations of a few minutes to several months. The creep data indicated changes in creep rate of up to 10 mm/year lasting for 6 months or more and a close similarity between the changes in creep rate on two adjacent creepmeters about 7 km apart. Earthquakes with magnitudes less than 4.0 do not appear to correspond in time to local changes in magnetic field greater than 0.75 γ or variations in the creep rate. There is no general correspondence between creep events and magnetic field variations. There is, however, an approximate correspondence, in both space and time, between the long-term changes in creep rate and the variations in magnetic field. In order to explain the observations presented in this study, it appears necessary to allow for a substantial amount of deep aseismic slip without any obvious attendant changes in the time distribution or size of the local earthquakes.

1. Introduction

Since early 1974 the U.S. Geological Survey has been measuring total magnetic field in central California with an array of proton precession magnetometers (Fig. 1). The purpose of these measurements is to identify local changes in the magnetic field that might be associated with the active faults in the region. The theory and laboratory studies which indicate that tectonomagnetic signals should occur as the result of stress induced changes in rock magnetization (piezomagnetic effect) have been reviewed a number of times previously (e.g., JOHNSTON *et al.*, 1976a; RIKITAKE, 1976; SMITH and JOHNSTON, 1976; STACEY and BANERJEE, 1974).

Most of the magnetometer stations are located in two areas within the San Andreas fault system: around the south end of the 1906 earthquake fault break, near station SN; and around the north end of the 1857 break, believed to be between

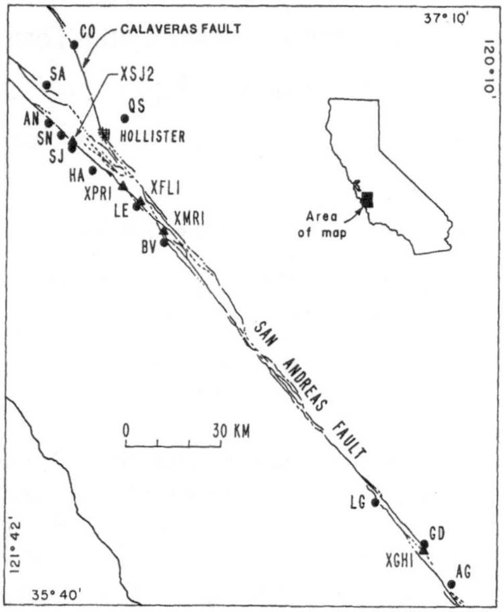

Fig. 1. Map of station locations and faults along the seismically active secton of the San Andreas fault in central California. Dots are magnetometer stations; triangles, creepmeter stations. Three magnetometer stations (not included in this study) are located northwest of the map area.

stations GD and AG (Fig. 1). The larger earthquakes in this region tend to occur near the ends of these fault breaks. Between these two areas the fault is presently creeping and the recorded earthquakes have magnitudes up to about 5.5. Near the center of the creeping section the rate of recorded creep is relatively high, about 30 mm/year, but it decays to zero near the two ends where strain accumulation might be expected (SAVAGE and BURFORD, 1971).

The magnetometer stations included in this study have detected local changes in the magnetic field with amplitudes as high as 1.8γ (JOHNSTON *et al.*, 1976a; SMITH and JOHNSTON, 1976). Although it is likely that these changes are produced by the piezomagnetic effect, they have not yet been observed to occur simultaneously with stress changes from local earthquakes with $M_L < 4.0$ or from surface creep events. An unanswered question concerns whether episodes of subsurface aseismic slip are the cause of these magnetic field changes.

The creepmeters record short-term creep events which typically last from a few minutes to a few days and have measured displacements of up to 5 mm. It is important to question whether these surface creep events reflect aseismic slip that extends down through the whole seismic zone ($\sim$15 km deep). Recent studies, using arrays of tiltmeters and strainmeters that have recorded near-simultaneous strain events associated with creep events (JOHNSTON *et al.*, 1976b; MCHUGH and JOHNSTON, 1976; MORTENSEN *et al.*, 1977), indicate that the creep seen at the surface probably occurs in the top 2 km of the crust. The short-term creep events, therefore, probably do not play a significant role in the release of accumulated strain in the seismic zone

(2 to 15 km deep). More likely they are just a surface response to deeper slip. Additional support for this possibility comes from continuous geodetic strain measurements, obtained near Hollister, over baselines 3 to 9 km long. These data indicate that accelerated slip episodes, lasting for several weeks and extending to a depth of about 10 km, precede surface creep events by several weeks (SLATER and BURFORD, 1978).

This paper will try to determine if deep aseismic slip can explain the observed local magnetic field changes by comparing magnetic, creep and earthquake data from the section of the San Andreas fault between the south end of the 1906 fault break and the north end of the 1857 break.

2. Data

The proton precession magnetometers used in this study have a sensitivity and precision of 0.25γ. All the magnetometer stations sample the total magnetic field simultaneously, within ± 0.2 sec, once a minute. The electronics are housed in an insulated fiberglass pit, 1.8 m deep, buried in the ground so that only the top few centimeters is exposed. The sensor is housed on top of a wooden post about 2 m above the ground. Each site is selected on the basis of low magnetic field gradient ($< 3\gamma$/m), remoteness from any cultural objects that could significantly contaminate the magnetic field, and proximity to known or suspected magnetic rocks adjacent to active faults. Each station automatically samples the total magnetic field, converts the data to a serial digital code, and telemeters the data to Menlo Park, California, via radio links and/or telephone lines. Each total field value is recorded on magnetic tape, which is later transferred to a computer where all data processing is done.

The diurnal variation of the total magnetic field, due to ionospheric and magnetospheric effects, is typically 40 to 60γ. Since tectonomagnetic signals are probably not more than a few gammas (STACEY and BANERJEE, 1974), it is necessary to reduce the diurnal variations by a factor of 50 or more. To accomplish this reduction, we first calculate the difference between two stations that are separated by less than a few tens of kilometers (usually adjacent stations). Simple differences reduce the diurnal variations by about a factor of 10. A 5-day average further reduces the diurnal variations to an amplitude of about 0.50 to 0.25γ. The magnetic data presented in this paper are 5-day running averages of differences between adjacent stations. These data show many variations with amplitudes of about 0.50γ. It is not clear what fraction of these variations are due to a tectonomagnetic source, ionospheric or magnetospheric effects, or some other physical process. For brevity we will discuss only those variations with amplitudes greater than 0.75γ.

The creep data used in this study are from the wire creepmeter network established by the U.S. Geological Survey (YAMASHITA and BURFORD, 1973). Five of these creepmeters were chosen for this study on the basis of nearness to magnetometer stations and completeness of data (Fig. 1). These data show that right-lateral fault creep occurs at an approximately constant rate for the time frame of this study. In

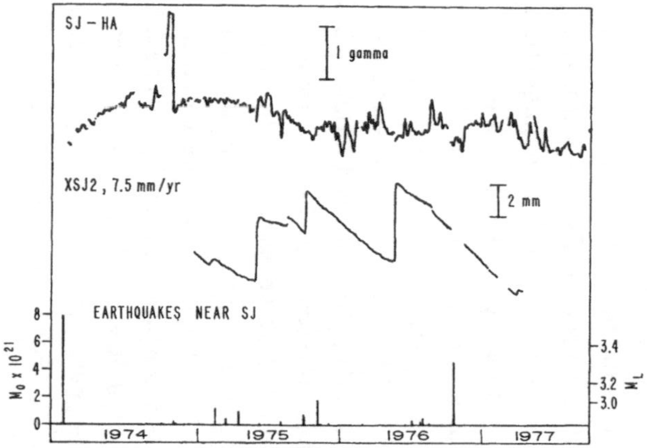

Fig. 2

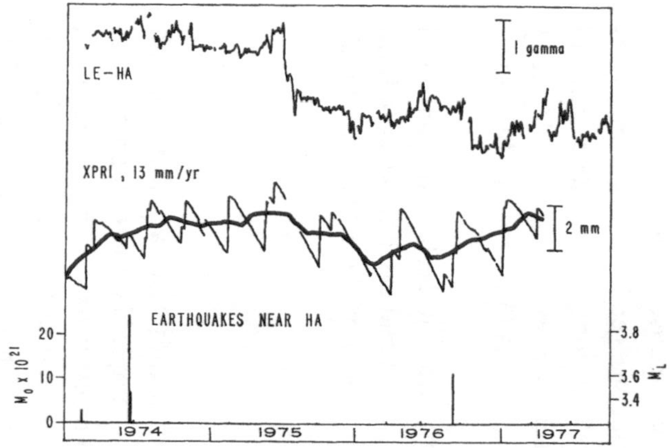

Fig. 3

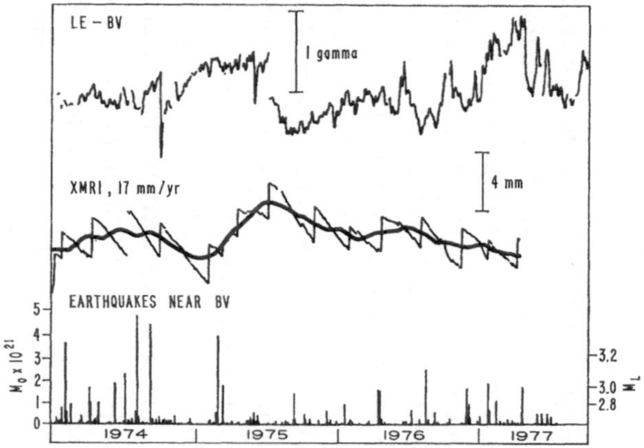

Fig. 4

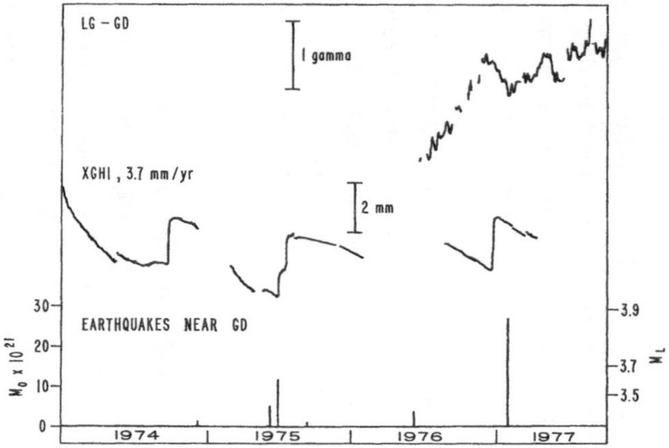

Fig. 5

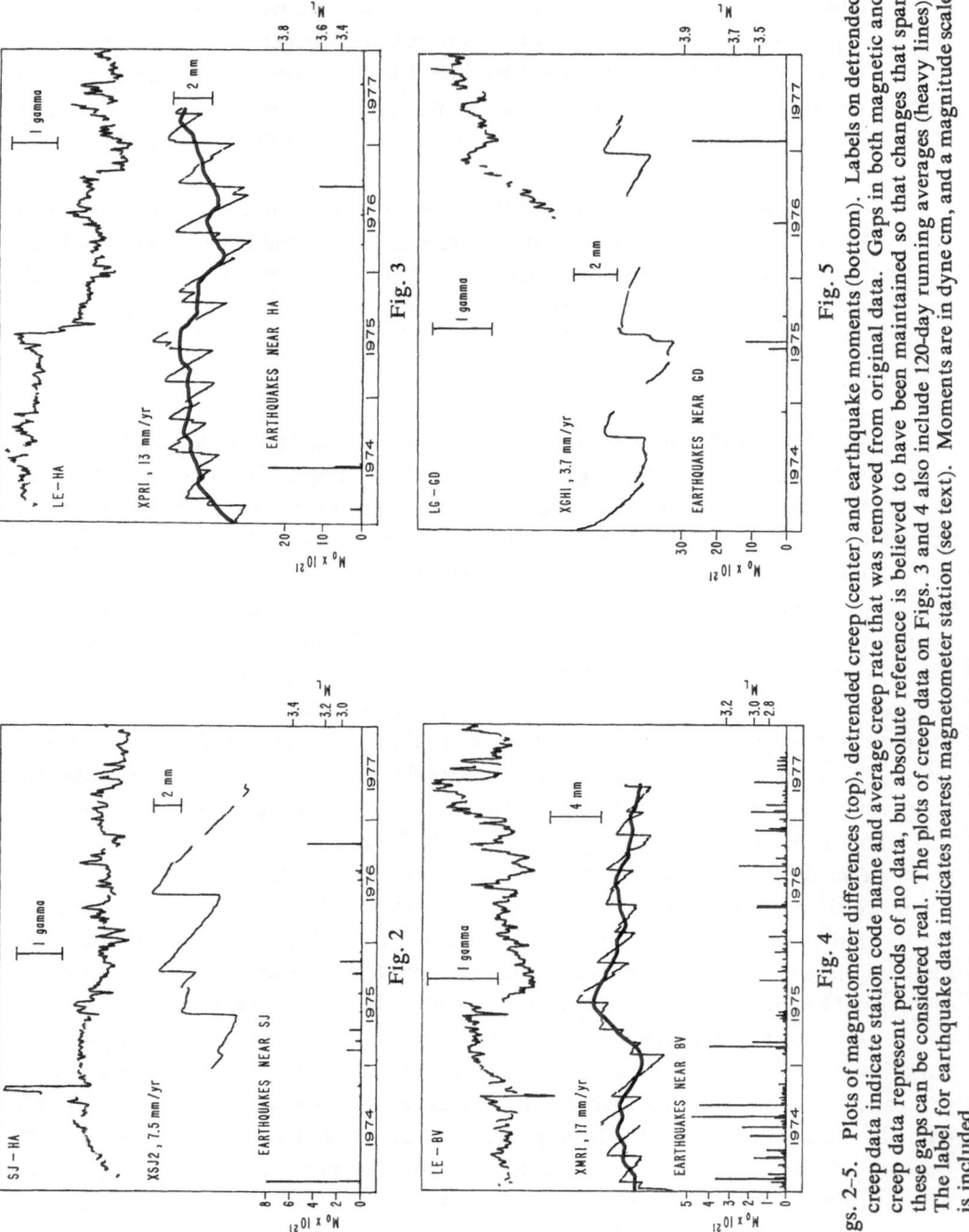

Figs. 2–5. Plots of magnetometer differences (top), detrended creep (center) and earthquake moments (bottom). Labels on detrended creep data indicate station code name and average creep rate that was removed from original data. Gaps in both magnetic and creep data represent periods of no data, but absolute reference is believed to have been maintained so that changes that span these gaps can be considered real. The plots of creep data on Figs. 3 and 4 also include 120-day running averages (heavy lines). The label for earthquake data indicates nearest magnetometer station (see text). Moments are in dyne cm, and a magnitude scale is included.

order to determine whether changes in creep rate correspond to changes in magnetic field, the average rate from these records has been removed by determining the best linear fit to the data and subtracting this line from the original values; a technique similar to that employed in a previous investigation of the relationship between creep rate and moderate earthquakes (BURFORD, 1976). The detrended creep data show the small variations of creep rate much more clearly than do the untreated data.

The earthquake data used in this study are from the unpublished earthquake catalog compiled by the U.S. Geological Survey. A lower magnitude cutoff of 1.3 was chosen because it is the lowest magnitude that is likely to provide a complete data set in the regions studied. The earthquakes were then chosen to include only those that appeared to have occurred on the San Andreas fault and within 8 km of a magnetometer station. When adjacent magnetometers were located closer than 16 km apart, the set of earthquakes was further divided to include only the earthquakes closest to each magnetometer station. The earthquake sets are denoted by 'earthquakes near (station code name)'. The largest earthquake included in this study has a magnitude of 3.9, and the rate of occurrence of earthquakes below magnitude 3.0 is fairly steady. The earthquakes show little variation in magnitude with time. Therefore, in order to show more clearly what may be the more relevant earthquakes in terms of association with tectonomagnetic signals, we have chosen to plot earthquake moments. Moments were calculated using a magnitude-moment relation derived from central California earthquakes (BAKUN *et al.*, 1976):

$$M_0 = 10^{(1.4M_L+17)} \text{ dyne cm}$$

where M_0 is the moment and M_L is local magnitude.

The comparison of magnetic data, creep data, and earthquakes was made by plotting all three of these parameters on the same time scale. For the time period and regions included in this study, local changes in the magnetic field greater than 0.75γ occur only on the following difference records: SJ-HA, LE-HA, LE-BV, and LG-GD. All four of these difference records are shown in Figs. 2, 3, 4, and 5, along with the most relevant plots of detrended creep data and earthquake moments. Gaps in the magnetic and creep data indicate periods where data were lost owing to a variety of reasons. In all cases we believe that the absolute reference has been maintained so that changes that span these gaps can be considered real.

3. Results

A significant change in magnetic field on the SJ-HA record occurred during October 1974, lasted for the whole month, and had a maximum amplitude of 1.8γ (Fig. 2). By comparison with other difference records, we know that this change was recorded at station SJ. Because the change was recorded only at station SJ, its source is probably within a few kilometers of this station. Unfortunately, the nearest creepmeter, XSJ2, was not operational until early November 1974. There are no obvious correlations between the XSJ2 creep data, the earthquakes near SJ, and the magnetic field record SJ-HA.

The local changes in magnetic field in October 1974 did occur one month prior to a magnitude 5.1 earthquake (SMITH and JOHNSTON, 1976). This earthquake is not shown in the plotted earthquake data because it occurred to the east of the San Andreas fault and about 11 km from station SJ. Because of the distance between the earthquake and the probable source of the magnetic field changes, and the lack of a magnetic signal at the time of the earthquake, there is probably no direct relation between them. However, the magnetic field change could be explained by aseismic slip on the San Andreas fault that may be related to the magnitude 5.1 earthquake by an interaction of the faults in the area, as discussed in detail by SMITH and JOHNSTON (1976).

The difference record LE-HA (Fig. 3) shows a magnetic signal with a maximum amplitude of 1.7 γ occurring during the period June to August, 1975. Comparison with other difference records indicates that signals were recorded at both LE and HA, but in the opposite sense so that the difference LE-HA has the effect of adding the changes from these two stations. The character of the changes recorded at HA and LE can be seen to be dissimilar by comparing the SJ-HA (Fig. 2) and LE-BV (Fig. 4) records. The change at HA is positive and mostly smooth while the change at LE is negative, starts with a sudden event, and then is smoother. The correspondence in time of these magnetic field changes suggests that they may have the same or a related source.

The creep data plotted with the LE-HA record are from station XPR1. The creep events at this station occur fairly regularly. With a 120-day running average (heavy line), it is possible to smooth these events and emphasize the long-term character of the creep data (Fig. 3). The average creep data show that only during the second half of 1975 was the creep rate significantly less than the average rate for the period of time included in this study. The start of the decrease in creep rate occurs during the change in magnetic field at stations LE and HA discussed above. The long-term trends of the averaged creep data from station XFL1 appear to be almost identical to XPR1, except that almost a year of data is missing starting from mid-1975. Both the XFL1 and the XPR1 data show that the creep rate at these stations was below average between mid-1975 and mid-1976, although the exact character of the change at XFL1 cannot be determined because of the missing data.

The plot of earthquakes near HA (Fig. 3) does not show any apparent correspondence to the LE-HA record. It is interesting to note that the larger earthquakes occur during times of higher than average creep rate. The plot of earthquakes near LE shows a fairly even rate of earthquakes up to magnitude 3.2 with no apparent correspondence to either the creep data or the LE-HA magnetic record.

There do appear to be some interesting correspondences between the LE-BV and XMR1 records. The difference record LE-BV (Fig. 4) shows three significant magnetic field changes. One change of short duration and 1 γ amplitude in September 1974 was recorded at station BV. The second change, occurring in mid-1975, was recorded at station LE and is the same signal discussed above in the section covering the LE-HA record. The third change, in late 1976 and early 1977, was one of complex character that appears to be recorded primarily at station LE. The detrended creep data from XMR1 have been smoothed with a 120-day running average, and both the original and smoothed data (heavy line) are plotted together in Fig. 4. The average rate of creep decreased slightly at the end of 1974 and then sharply increased by 10 mm/year during the first half of 1975. The long-term trends on both records are approximately mirror images of each other. The two short-term changes in October 1974 and June 1975 approximately bracket in time the high rate of creep during the first half of 1975 and also occur within one day of creep events. (Because of telemetry failures, the exact time and duration of these changes in magnetic field cannot be determined.) However, all but the October 1974 signal on the LE-BV record were recorded at station LE, 11 km northwest of creepmeter XMR1. There does not appear to be any significant correspondence between the earthquakes near BV and the LE-BV record or the XMR1 record.

The LG-GD record shows a 1.5γ change in magnetic field during the second half of 1976 (Fig. 5). This change was recorded at station GD. The first data from stations LG and GD were obtained in June 1976, so that the time when this change began is not known. A creep event at XGH1 and a magnitude 3.9 earthquake, the largest to occur within 8 km of station GD during the recording period, occurred just after the end of the change in magnetic field (Fig. 5).

4. Discussion

The data presented in this study indicate that small-magnitude earthquakes do not correspond in time with the changes in either the magnetic field or the creep rate. This fact is not too surprising since earthquakes in California of the size included in this study ($1.3 < M_L < 3.9$) appear to have slip dimensions of less than 1 km (Bakun *et al.*, 1976), stress drops of less than 20 bars, and depths of between 5 and 10 km (Thatcher and Hanks, 1973; Wesson *et al.*, 1973). Assuming a magnetic susceptibility of 10^{-3} emu, tectonomagnetic models (Johnston, 1978) using the above parameters show that these earthquakes would not, by themselves, generate a surface anomaly greater than 0.1γ. Of course, if the earthquake is accompanied or triggered by larger scale readjustments of regional stress, then these could be reflected in the magnetic data. These stress changes could result, for example, from related aseismic slip on the fault (Stuart and Johnston, 1974) or be a consequence of the initial stress conditions near the fault. These conditions might be modified by earthquakes or other fault behavior such as pore pressure changes, comminution, and chemical changes. In either case, the changes in magnetic field would not necessarily be expected to occur at the same time as earthquakes.

Another general conclusion evident in these data and discussed in more detail by Johnston *et al.* (1978) is that the large majority of short-term creep events do not correspond to changes in magnetic field. Two possible exceptions, where creep events occur within one day of short-term changes in magnetic field, are evident on the LE-BV record (Fig. 4) in October 1974 and July 1975. Although the changes in magnetic field are not likely to result from the surface creep events, it is possible that deep aseismic slip occurred at about the same time and that this deep slip is the source of both the changes in magnetic field and the creep events. It is interesting that these two creep events occur near the beginning and end of the accelerated creep rate during the first half of 1975. Since only two out of several dozen creep events correspond in time to changes in magnetic field, it is certainly possible that the correspondence is a coincidence.

The most exciting result is the approximate correspondence, in both space and time, of changes in local magnetic field and long-term creep rate. However, the data are too sparse and the time span too short to determine the significance of these correspondences. If the magnetic changes do reflect crustal stress changes in the region and are related to changes in long-term creep rate, than substantial energy transference is apparently occurring aseismically at depths below 2 km.

Models of spatially varying aseismic slip at depths between 2 and 10 km can be fit to the creep data and used to generate tectonomagnetic models that satisfy the magnetic data. These models will not be proposed in detail here since, without additional deformation measurements, it is not possible to demonstrate independently the existence of these failure patches or to constrain the parameters of these models. The correspondence in time between the long-term creep rate at XMR1 and the changes in magnetic field recorded at LE, 11 km to the northwest, is perhaps a good example of how these models might work. If a patch of the fault were slipping, stress concentration would be expected at the edges of this patch. If the patch is centered at XMR1 and extends northwest to a point near LE, then it would not be unreasonable to expect a change in creep rate at XMR1 to be related to a magnetic field change at LE. As more data become available, it will be possible to test and extend these models and their implications further. The occurrence of a moderate magnitude earthquake that breaks a substantial part of the seismic zone will also provide critical data on the amplitudes and interrelationship of the magnetic, creep, seismic and other data.

It is important to question whether any of the general fault models are precluded by these observations. Regardless of the details, the simplest general models of the fault in which the slip is assumed to be uniform or where slip occurs only during earthquakes, are certainly not consistent with these data. The intriguing patterns of behavior at, for example, GD, when a creep event and a magnitude 3.9 earthquake occurred shortly after the end of a change in magnetic field (Fig. 5), and at LE and HA in 1975 during periods of retarded creep on nearby XPR1 and XFL1 creepmeters and of accelerated creep on XMR1 (Figs. 3 and 4), argue for more complex and heterogeneous fault mechanics.

5. Conclusions

1) Local variations in magnetic field with amplitudes as high as 1.8γ occur within the seismically active segment of the San Andreas fault between the south end of the 1906 earthquake fault break and the north end of the 1857 break.

2) Long-term detrended creepmeter records along this section of the San Andreas fault show significant changes in creep rate (up to 10mm/year) lasting for several months.

3) Earthquakes with magnitude less than 4.0 do not appear to correspond in time to local changes in magnetic field greater than 0.75γ or long-term variations in creep rate.

4) In general, the short-term creep events do not correspond with local changes in magnetic field greater than 0.75γ or the local earthquakes.

5) The long-term changes in creep rate show an approximate correspondence in time and space to some long-term changes in magnetic field. The data are too sparse to determine the significance of these apparent correspondences.

6) For fault models to explain the observations presented in this study, it appears necessary to allow for a substantial amount of deep aseismic slip without any obvious attendant changes in the time distribution or size of the local earthquakes.

REFERENCES

BAKUN, W.K., C.G. BUFE, and R.M. STEWART, Body wave spectra of central California earthquakes, *Bull. Seismol. Soc. Am.*, **66**, 363–384, 1976.

BURFORD, R.O., Fluctuations in rates of fault creep associated with moderate earthquakes along the central San Andreas fault, *EOS, Trans. Am. Geophys. Union*, **57**, 1012, 1976.

JOHNSTON, M.J.S., Local magnetic field variations and stress changes near a slip discontinuity on the San Andreas fault, *J. Geomag. Geoelectr.*, **30**, 511–522, 1978.

JOHNSTON, M.J.S., B.E. SMITH, and R.J. MUELLER, Tectonic experiments and observations in western U.S.A., *J. Geomag. Geoelectr.*, **28**, 85–97, 1976a.

JOHNSTON, M.J.S., S. MCHUGH, and R.O. BURFORD, On simultaneous tilt and creep observations on the San Andreas fault, *Nature*, **260**, 691–693, 1976b.

JOHNSTON, M.J.S., B.E. SMITH, and R.O. BURFORD, Local magnetic field measurements and fault creep observations on the San Andreas fault, *Tectonophysics*, 1978 (in press).

MCHUGH, S. and M. JOHNSTON, Short period nonseismic tilt perturbations and their relation to episodic slip on the San Andreas fault, *J. Geophys. Res.*, **81**, 6341–6345, 1976.

MORTENSEN, C.E., R.C. LEE, and R.O. BURFORD, Simultaneous tilt, strain, creep, and water level observations at the Cienega Winery south of Hollister, California, *Bull. Seismol. Soc. Am.*, **67**, 641–650, 1977.

RIKITAKE, T., *Earthquake Prediction*, pp. 197–211, Elsevier, Amsterdam, 1976.

SAVAGE, J.C. and R.O. BURFORD, Discussion of paper by C.H. Scholz and T.J. Fitch, Strain accumulation along the San Andreas fault, *J. Geophys. Res.*, **86**, 6469–6479, 1971.

SLATER, L.E. and R.O. BURFORD, A comparison of long-baseline strain data and fault creep records obtained near Hollister, California, *Tectonophysics*, 1978 (in press).

SMITH, B.E. and M.J.S. JOHNSTON, A tectonomagnetic effect observed before a magnitude 5.2 earthquake near Hollister, California, *J. Geophys. Res.*, **81**, 3556–3560, 1976.

STACEY, F.D. and S.K. BANERJEE, *The Physical Principles of Rock Magnetization*, pp. 146–155, Elsevier, Amsterdam, 1974.

STUART, W.D. and M.J.S. JOHNSTON, Tectonic implications of anomalous tilt before central California earthquakes, *EOS, Trans. Am. Geophys. Union*, **56**, 1196, 1974.

TALWANI, P. and R.L. KOVACH, Geomagnetic observations and fault creep in California, *Tectonophysics*, **6**, 69–73, 1972.

THATCHER, W. and T.C. HANKS, Source parameters of southern California earthquakes, *J. Geophys. Res.*, **78**, 8547–8576, 1973.

WESSON, R.L., R.O. BURFORD, and W.L. ELLSWORTH, Relationship between seismicity, fault creep and crustal loading along the central San Andreas fault, in *Proceedings of the Conference on Tectonic Problems of the San Andreas Fault System*, edited by R.L. Kovach and A. Nur, pp. 303–321, School of Earth Sciences, Stanford University, Stanford, California, 1973.

YAMASHITA, P.A. and R.O. BURFORD, Catalog of preliminary results from an 18-station creepmeter network along the San Andreas fault system in central California for the time interval June 1969 to June 1973, *U.S. Geol. Surv. Open File Report*, 215 pp., 1973.

Geomagnetic Induction Study of the Seismically Active Fault along the Southwestern Coast of the Sea of Japan

Junichiro MIYAKOSHI* and Akira SUZUKI**

*Institute of Earth Science, Tottori University, Tottori, 680 Japan
**Geophysical Institute, Kyoto University, Kyoto, 606 Japan

(Received November 5, 1977)

Geomagnetic and telluric current observations were conducted on and around the seismically active fault, Yoshioka-Shikano Fault, from 1976 to 1977.

The confined plane of the geomagnetic variation expressed as $\Delta Z = A\Delta X + B\Delta Y$, was calculated for each station by applying the transfer-function techniques or the least-square method to the records of Pi 1–2 type pulsations.

It was found that the landward increasing tendency of the A values of the confined planes caused by the electromagnetic coastal effect of the Sea of Japan, was slightly interrupted on the northern edge of the fault and also, that the amplitude of the N-S component of geomagnetic variation ($\Delta \dot{X}$) was considerably enhanced upon the fault. These results are proposed to be due to the effect of the swelling of the electrically conducting medium in the crust beneath the seismically active fault.

A tentative analysis was also made for the telluric current records observed on the fault, comparing it with the data obtained at the same site just after the Tottori Earthquake. As for the direction of polarization of the electric field there was no noticeable difference between them.

1. Introduction

Geomagnetic and telluric current observations were conducted on and around the seismically active fault, Yoshioka-Shikano Fault, from 1976 to 1977. The aim of these observations is to study the geoelectric structure of the crust and upper mantle beneath the seismically active fault, and also, to see if the time variation of the geoelectric characteristics of the epicentral region relating to the earthquake occurrence might exist.

The map of the studied area, which shows the locations of the fault, observation sites and bathymetric contours is given in Fig. 1. The Yoshioka-Shikano Fault, of strike-slip type, is estimated to run for about 30 km in length in an almost east-west direction along the southwestern coast of the Sea of Japan (KANAMORI, 1972). This fault was formed at the time of the Tottori Earthquake ($M=7.4$) which occurred in 1943. Crustal displacement in the vicinity of the fault is shown in Fig. 2. A number of microearthquakes have been observed to occur within a narrow width, at a depth less than about 20 km beneath the fault up to the present time (TSUKUDA *et al.*, 1976) (Fig. 3).

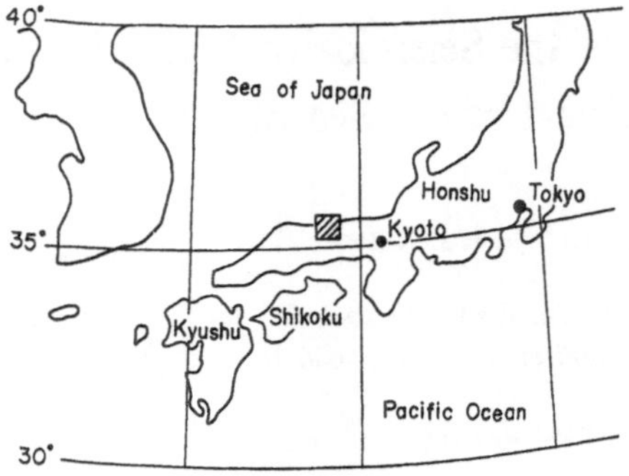

Fig. 1a. Location of the studied area.

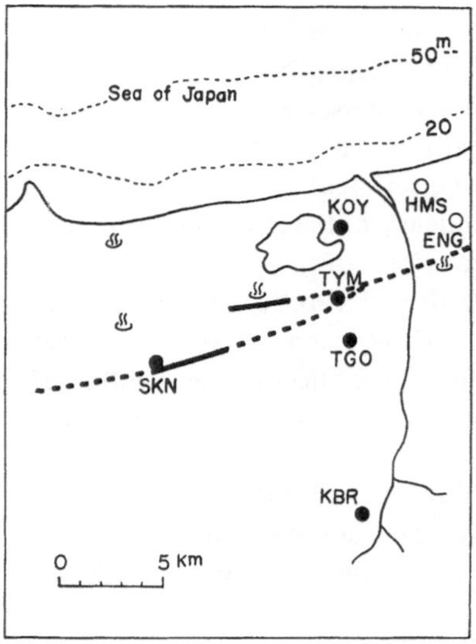

Fig. 1b. Enlarged map of the studied area. Yoshioka-Shikano Fault, the observation sites and the bathymetric contours are also shown.

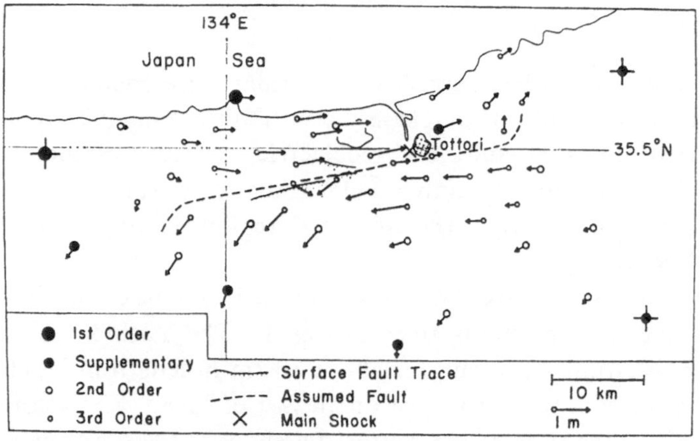

Fig. 2. The displacements of the triangulation points during the period from 1891 to 1957 (after SATO and NAKANE, 1972).

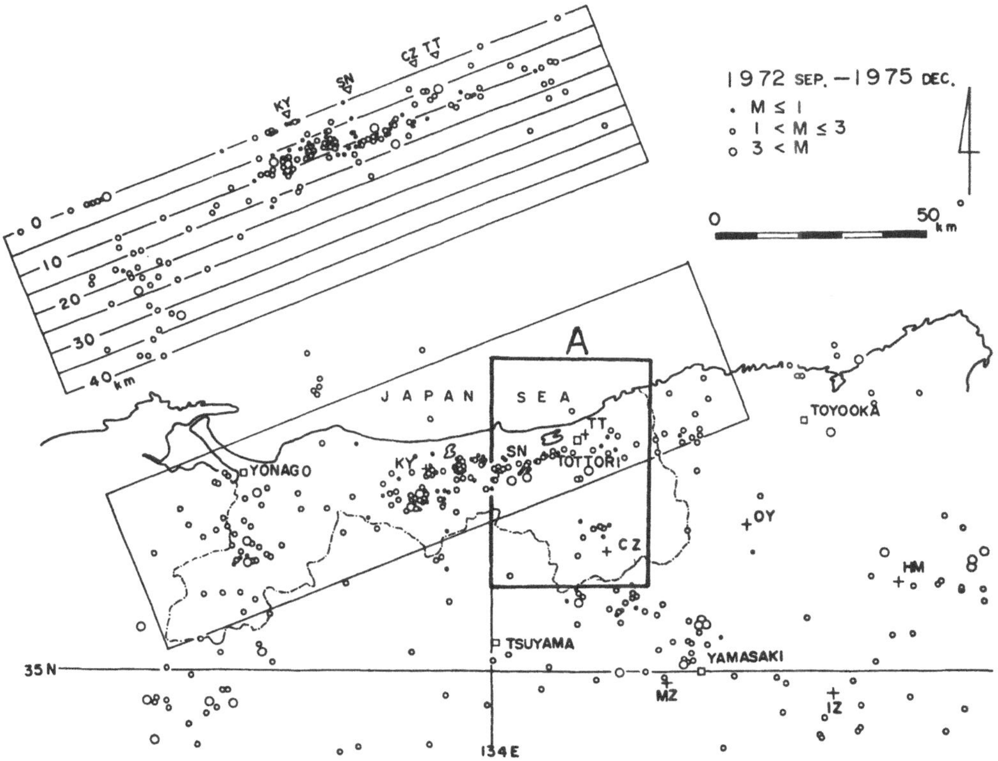

Fig. 3. Hypocenters distribution (after TSUKUDA, 1976). A: Studied area in this paper.

It was pointed out, recently, that the groundwater plays an important role in the occurrence of shallow earthquakes (ANDERSON and WHITCOMB, 1975). Meanwhile, the water content of the rocks is recognized to have a remarkable effect to their electrical conductivity (PARKHOMENKO, 1967).

It is the objective of this paper, from the point of view mentioned above, to study the electrical conductivity structure beneath the seismically active fault and also, to investigate if the time variation of the geoelectric character of the fault might exist in relation to earthquake occurrences.

2. Geoelectric Sounding of the Fault

Before starting the observations, a geoelectric sounding was made to clarify the shallow structure of the fault. Geoelectric soundings were carried out applying the Schlumberger electrodes spacing method to the Yoshioka Fault. Figures 4a and 4b show obtained apparent resistivity curves of vertical sounding and lateral sounding, respectively. From Fig. 4a and from the data obtained in geological studies, we inferred that the thickness of the sediment covering the basement rock of granite is less than 100 m in this area and that the electrical resistivity gradually decreases with depth. Low resistivity values seen on the lateral sounding curve are considered as indicating the existence of fractured zones saturated with groundwater.

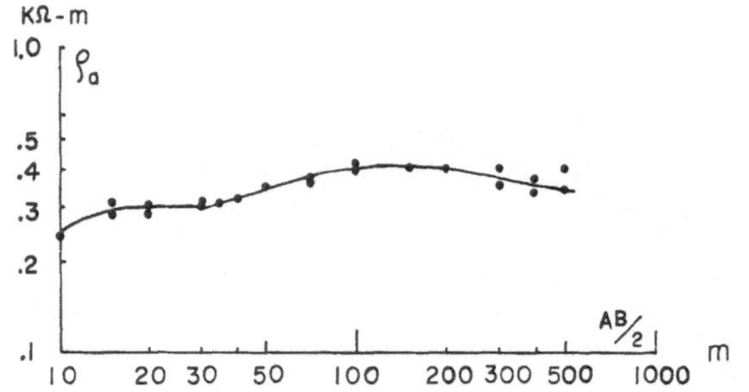

Fig. 4a. Vertical electrical sounding curve of the Yoshioka Fault.

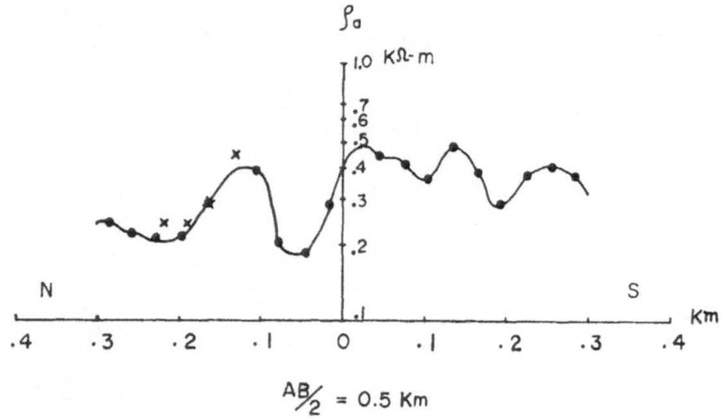

Fig. 4b. Lateral electrical sounding curve of the Yoshioka Fault.

3. Geomagnetic Observations and Their Results

The CA Group (Conductivity Anomaly Research Group) has previously shown that the depth to the electrically high conductive layer situated in the lower crust or upper mantle must be rather shallow in this region. They estimated the depth to the surface of the conductor, as 30 km or less (MIYAKOSHI, 1976). Meanwhile, the epicenters of the main shock and the aftershocks of the Tottori Earthquake and also the microearthquakes in this region are all found to be distributed within a depth of about 20 km (OIKE, 1975).

Taking these geophysical data and the skin depth of the inducing magnetic field into account, we decided to make an observation of the Pi 1–2 pulsations using the induction-magnetometers for the geomagnetic induction study of this region. The frequency characteristics of the magnetometers are shown in Fig. 5. Observation sites were chosen to distribute in the north-south direction crossing perpendicular to the fault and to the coast of the Sea of Japan. The observations were made mainly in the nighttime to avoid artificial disturbances. In Fig. 6 examples of the records are shown.

The confined plane of the geomagnetic variation is expressed as $\Delta Z = A(f)\Delta X + B(f)\Delta Y$, where $A(f)$ and $B(f)$ are the functions of the frequency of the geomagnetic variation. This confined plane is almost parallel to the surface of the adjacent electrically high conductive medium such as the sea water or mantle. Transfer-function technique or least square method or both were employed to calculate the values A

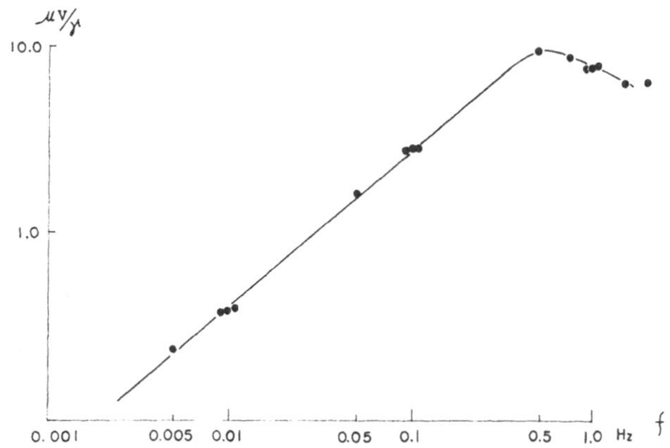

Fig. 5. Frequency characteristics of the induction-magnetometers used in this study.

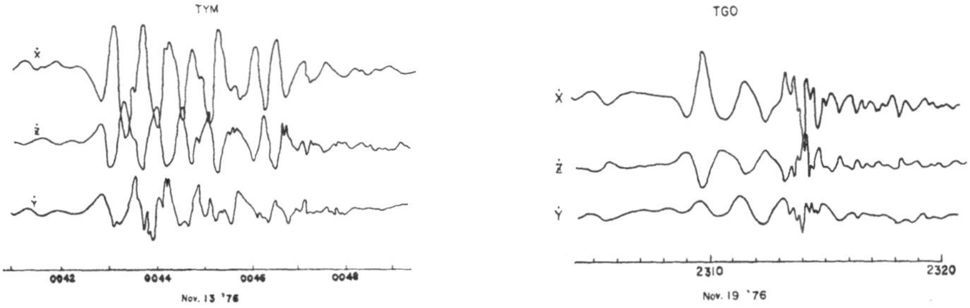

(a) (b)

Fig. 6a, b. Examples of the magnetograms. The ordinate scale is arbitrary.

and *B*, considering the characteristics of the records. The calculation of transfer functions were made using the program written by Y. Honkura (HONKURA, 1971). As *B* values are quite small for all observation sites, the *A* values are preferred for discussion in this paper. Examples of the calculated transfer function and the correlation between $\Delta\dot{Z}/\Delta\dot{Y}$ and $\Delta\dot{X}/\Delta\dot{Y}$ are shown in Fig. 7. The geographical distribution of *A* values thus obtained are shown in Fig. 8. It was found in the calculation of the *A* values for Hamasaka (HMS) and Engoji (ENG) that the calculated values by the least square method are less by 0.15 than that calculated by the technique of the transfer function. This discrepancy might be caused by the out-of-phase part of the *Z* component which was not separated in the calculation by the least square method. The corrected *A* values, adding by 0.15 to the values calculated by the least square method, were adopted to construct the curve of the geographical distribution of *A* values in Fig. 8.

Recently, T. Rikitake pointed out that the observation of the horizontal component of the short-period geomagnetic variations might be also an advantageous strategy to find the underground electrical conductivity heterogeneity and its time variation of the small scale (RIKITAKE, 1976).

Bearing this proposal in mind, we made the observations of the *X* component at each site successively, always synchronized with Koyama site. In Fig. 9 are plotted

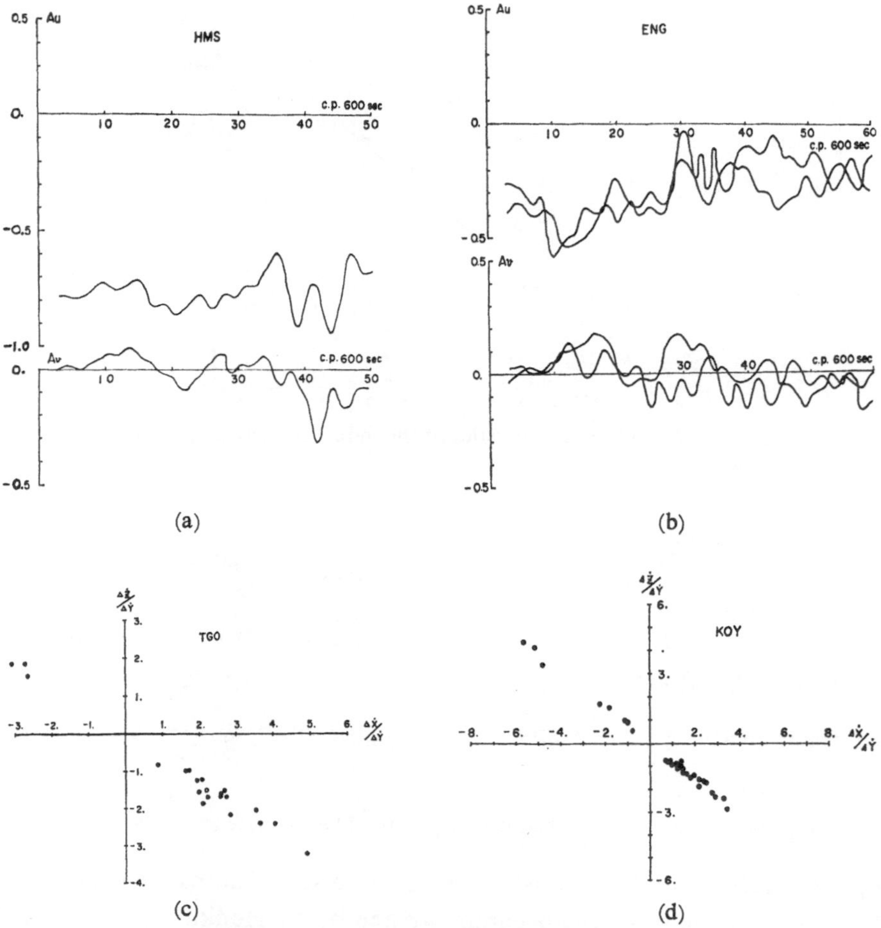

Fig. 7a, b. Examples of the transfer functions for the Pi 1–2 pulsations. *Au*, real part of *A*; *Av*, imaginary part of *A*.

Fig. 7c, d. Examples of the correlation between $\Delta\dot{Z}/\Delta\dot{Y}$ and $\Delta\dot{X}/\Delta\dot{Y}$ for the Pi 1–2 pulsations.

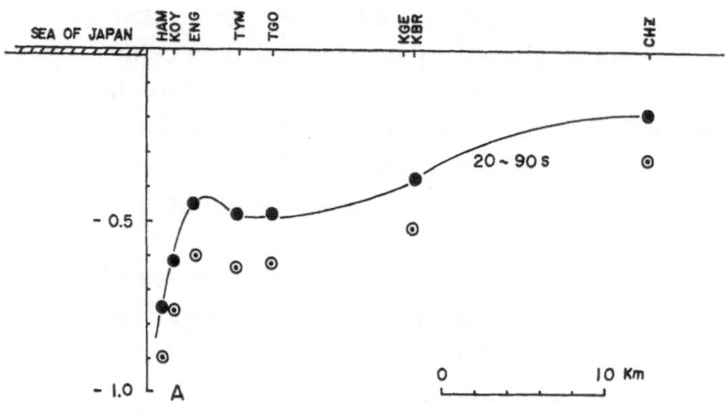

Fig. 8. Profile of the calculated *A* values for the Pi 1–2 pulsations, along a line perpendicular to the fault and the coast line.

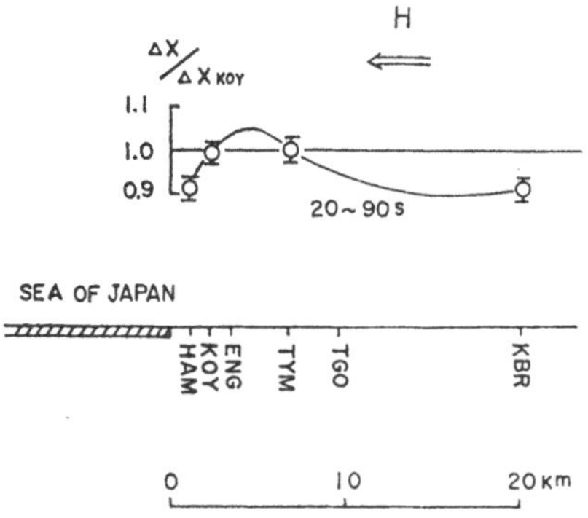

Fig. 9. Profile of the field ratio of the X component relative to that at Koyama for the Pi 1-2 pulsations, along a line perpendicular to the fault and to the coast line.

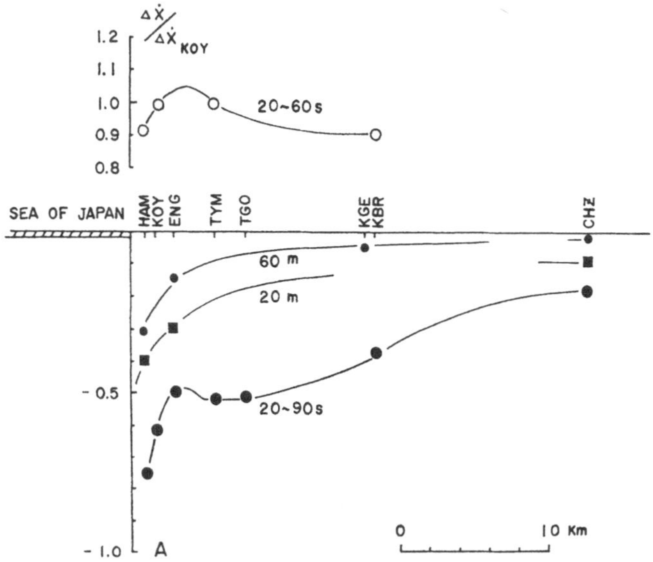

Fig. 10. Profile of the revised A values and $\Delta\dot{X}/\Delta\dot{X}_{KOY}$ along a line perpendicular to the fault and to the coast line.

the observed amplitude ratio of X component for each site relative to the simultaneously observed amplitude at the Koyama site. Small but evident enhancement of the amplitude of the variation are observed at the sites situated above or close to the fault. A values (real part of the $\Delta\dot{Z}/\Delta\dot{X}$) of these sites are, therefore, corrected slightly taking the spatial difference of $\Delta\dot{X}$ into account. The resultant geographical distribution of the revised A values together with the amplitude ratio of X component are shown in Fig. 10. In Fig. 10, the A values for the longer period variation previously obtained by the CA Group using the fluxgate-magnetometers, are also included. The tendency for all the A values of the various periods to decrease towards the seacoast is clearly seen in Fig. 10. It is interesting, on the other hand, to see that the geographical distribution of A values for the Pi 1–2 pulsations (20 to 90 sec in this case) has a remarkable inflection at around the northern edge of the fault, near to the place of the maximum of the amplitude of $\Delta\dot{X}$.

4. Earth Current Observation on the Fault

Earth current observations were made at Shikano (SKN) situated on the Shikano Fault. The observation site is located at the same place where the earth current observation was made just after the Tottori Earthquake (NAGATA, 1944).

Electrodes settings are almost parallel to the fault line and 242.5 m in distance for the geomagnetic E-W component, which is perpendicular to the fault and 233.8 m in distance for the geomagnetic N-S component. Electrodes used are the copper-copper sulfate electrodes. Figure 11 is an example of the records. As clearly seen in this figure, the variation is dominantly in the north-south component, and the vectors of variation are almost all confined in the direction 20° eastward from the north. This direction of polarization is similar to that observed just after the Tottori Earthquake in 1943. The impedance tensors expressed as

$$\Delta E_X = Z_{11}\Delta X + Z_{12}\Delta Y\,, \qquad E_Y = Z_{21}\Delta X + Z_{22}\Delta Y$$

were calculated by the least square method for the variation of about a 30-min period and shown in Table 1.

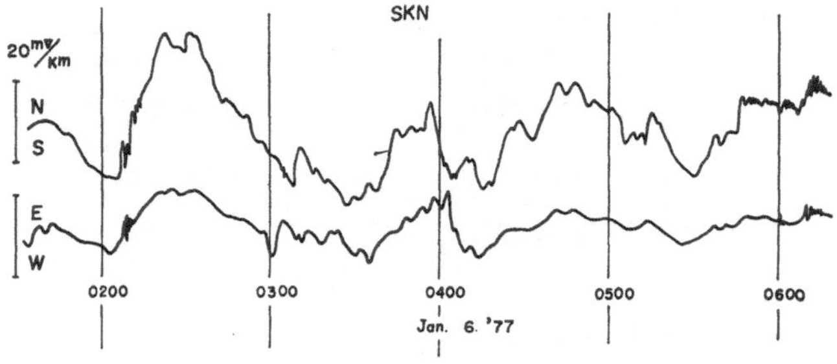

Fig. 11. Example of the earth-current records observed on the Shikano Fault.

Table 1. Impedance tensors calculated for about 30-min period variations in the earth-current records observed on the Shikano Fault.

Z_{11}	-1.82 ± 0.38 mV/km γ
Z_{12}	1.63 ± 0.53
Z_{21}	0.11 ± 0.34
Z_{22}	1.01 ± 0.15

As there was no seismic activity during the period of observation, we could not find any correlation between the time variation of the characteristics of telluric current and seismic activity.

5. Discussion

In Fig. 10, the decreasing trend to the seaward is clearly observed for all geographical distribution of A values of the various periods. Generally speaking, this is ascribed to the coastal effect of the Sea of Japan.

Rather small A values on the seacoast and rapid change on the landward side, which are remarkable for the 20- and 60-min periods, suggest the existence of a coupling effect between the electric current systems that are induced in the seawater and in the underground electrically high conducting layer, situated in the lower crust or in the upper mantle beneath the Sea of Japan. The surface of the highly conducting layer is inferred to extend from beneath the southwestern portion of the Sea of Japan to this region at the depth of about 30 km (MIYAKOSHI, 1976).

It should be noted, on the other hand, that the geographical distribution of A values for the Pi 1–2 pulsations (20 to 90 sec in this case) have evident inflection on the northern edge of the fault. Besides, the enhancement of the amplitude of $\Delta\dot{X}$ are clearly observed for the sites situated immediately above the fault.

These results strongly suggest the existence of a highly conducting medium beneath the fault, extending into the crust from the surface of conductive stratum above mentioned.

The underground electrical conductivity structure beneath the fault inferred from the overall results of observations is shown schematically in Fig. 12. An upraised highly conductive medium of prismatic shape having a triangular cross-section

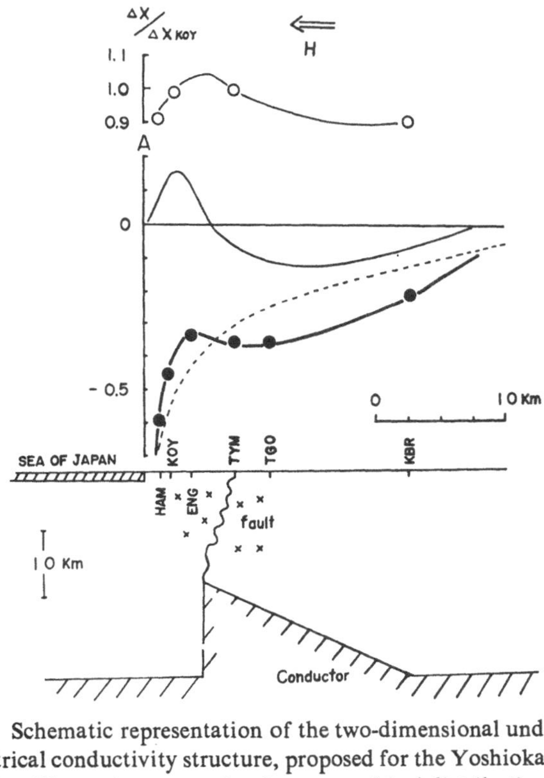

Fig. 12. Schematic representation of the two-dimensional underground electrical conductivity structure, proposed for the Yoshioka-Shikano Fault. Illustrative curves for the geographical distribution of the A values are also shown.

is proposed. Figure 12 is a two-dimensional representation in the north-south profile perpendicular to the fault and to the seacoast. The axis of the prism coincides with the strike of the fault. The model is the same in shape as one of the models which was calculated by T. Rikitake and K. Whitham to interpret the geomagnetic variation anomaly observed in Alert, Canada (RIKITAKE and WHITHAM, 1964). The model has been chosen taking the tendency of the spatial variation of *A* values for the Pi 1–2 pulsations into account, that is, the rapid change at the northern edge on the fault and the gradual change towards inland.

In Fig. 12, the schematic illustration for the geographical distributions of *A* value are also indicated. The dotted curve means the distribution caused by the coastal effect and the thin solid curve shows the effect caused by the prismatic conductor with the triangular cross-section. Combining the two effects described, a thick solid curve was obtained which models the observed distribution of *A* values. The depth to the top of the conductor is estimated as approximately 20 km, assuming the surface conductivity as 0.1 kΩ m and the period of the inducing field as 20 sec.

A possible cause of the upraised electrically conducting medium is the introduction of groundwater into the brittle region beneath the seismically active fault. This region might play the role of a water reservoir supplying the groundwater to the hypocentral regions of the microearthquakes where dilatancy processes are perhaps taking place. The result, shown in Fig. 13, indicates that the Poisson's ratio of the crust in this region falls off abruptly at a depth of 15 km (HASIZUME, 1970), and is consistent with our inference.

Although other exact tectonic evidence to corroborate the inferred electrical conductor beneath the Yoshioka-Shikano Fault, is not known, and there are few observational data to support a quantitative discussion at present, it must be emphasized that, as the vertical electrical sounding and boring data demonstrated, the thickness of the sediments is so thin in this region (about 20 to 40 m in Yoshioka and its vicinity) that the magnetic field from the induced electric current in the sediments is considered to be negligible.

The results of the earth-current observations carried out on the Shikano Fault, demonstrate that the direction of polarization of the electric field is quite similar to that observed just after the Tottori Earthquake in 1943. This might be considered as an indication of the stability of the direction of polarization.

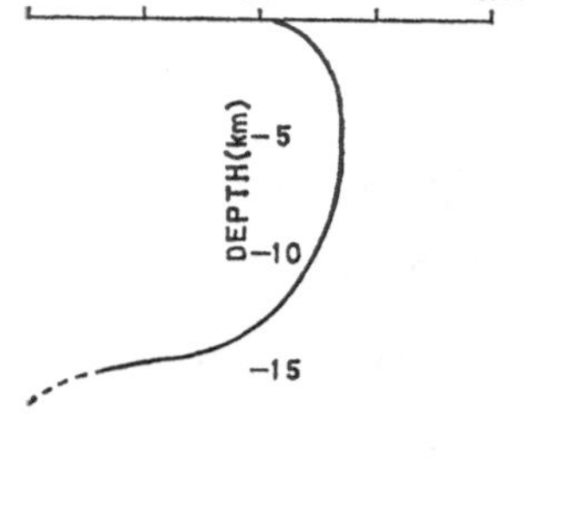

Fig. 13. Poisson's ratio variation with depth in the crust obtained in the study of microearthquakes (after HASHIZUME, 1970).

Although many questions are left for further study, the following results of this experimental study emerge:

1) Small enhancement of the amplitude of the N-S component of the geomagnetic variation and an inflection on the coastal effect curve of A value of the geomagnetic variation were observed for the Pi 1–2 events on and around the seismically active Yoshioka-Shikano Fault running almost in the E-W direction along the coast of the Sea of Japan.

2) They might be explained as a combined geomagnetic induction phenomenon caused by the coastal effect and by the upraising of the electrically high conducting medium underneath the fault.

3) The shape of the conductor is tentatively inferred to be prismatic with the axis underneath the fault and with a triangular cross-section.

4) The depth to the surface of the upraised conductor is likely to be about 20 km. This depth may have some connection with the depth at which the abrupt change of the Poisson's ratio takes place and with the base of the focal depth distribution of the microearthquakes in the fault zone.

5) The cause of the conductor upraising may be explained as the result of the water intrusion into the cracked zone in the fault.

6) The direction of polarization of the electric field observed on the Shikano Fault is 8° eastward from north, and this is the same as that observed just after the Tottori Earthquake occurred in 1943.

The authors want to express their hearty thanks to the principals and teachers who kindly permitted them to use a part of the buildings and campus of the schools for observations. The authors are also much obliged to the members of the CA Group for furnishing them the data observed by fluxgate-magnetometers. They are much indebted to Miss E. Inamura for her assistance in carrying out the computation.

The expense of this study was partly supported by a grant from the Ministry of Education. The computation of the transfer functions was done by a FACOM 230-60 in the Computing Center, Kyoto University.

REFERENCES

ANDERSON, D.L. and J.H. WHITCOMB, Time-dependent seismology, *J. Geophys. Res.*, **80**, 1497–1503, 1975.

HASHIZUME, M., On the nature of the crust—Investigation of microearthquake, *Bull. Disaster Prev. Res. Inst., Kyoto Univ.*, **20**, 53–64, 1970.

HONKURA, Y., Geomagnetic variation anomaly on Miyake-jima Island, *J. Geomag. Geoelectr.*, **23**, 307–333, 1971.

KANAMORI, H., Determination of effective tectonic stress associated with earthquake faulting, the Tottori earthquake of 1943, *Phys. Earth Planet. Inter.*, **5**, 426–434, 1972.

MIYAKOSHI, J., Electrical conductivity structure beneath the Japan Island Arc by geomagnetic induction study, *Tech. Rep. Geol. Surv.*, 31–40, 1976 (in Japanese).

NAGATA, T., Variation in earth-current in the vicinity of the Shikano-fault, *Bull. Earthq. Res. Inst., Tokyo Univ.*, 72–82, 1944 (in Japanese).

OIKE, K., On a list of hypocenters compiled by the Tottori Microearthquake Observatory, *Zisin*, **28**, 331–346, 1975 (in Japanese).

PARKHOMENKO, E.I., *Electrical Properties of Rocks*, pp. 119–128, Plenum Press, New York, 1967.

RIKITAKE, T., Crustal dilatancy and geomagnetic variations of short period, *J. Geomag. Geoelectr.*, **28**, 145–156, 1976.

RIKITAKE, T. and K. WHITHAM, Interpretation of the Alert anomaly in geomagnetic variation, *Can. J. Earth Sci.*, **1**, 35–62, 1964.

TSUKUDA, T., S. NAKAO, and Y. KISHIMOTO, Recent seismicity in the Tottori area, *Bull. Disaster Prev. Res. Inst., Kyoto Univ.*, **19**, 1–12, 1976 (in Japanese).

Time Dependence of Magnetotelluric Fields in a Tectonically Active Region in Eastern Canada

R.D. KURTZ and E.R. NIBLETT

Earth Physics Branch, Department of Energy, Mines and Resources, Ottawa, Canada

(Received November 3, 1977)

Magnetotelluric fields have been monitored for 3 years near the centre of seismicity in a tectonically active region on the north shore of the St. Lawrence River. The results indicate that electrical properties of upper crustal layers are strongly time-dependent in this area, and changes of more than 30% in the impedance tensor have been detected over a period of a few months. However, the most important part of the measured time dependence appears to be a trend increase in impedance of about 14% per year. There have been only two earthquakes greater than magnitude 3.0 in the area since recording began in 1974 and it has not been possible to develop a clear association between seismic activity and resistivity changes. Seasonal variations in the temperature and salinity of the nearby St. Lawrence River may be a contributing factor. Much less change in impedance was observed at similar MT recording stations located outside the zone of seismicity.

1. Introduction

In the Province of Quebec the triangular region between the Saguenay and St. Lawrence Rivers and bounded on the west by a line joining Quebec City and Lac Saint-Jean (Fig. 1) has exhibited anomalous and unusual geophysical behaviour. This part of Quebec is generally referred to as the Laurentide region. Some notable aspects of its character are high elevation accompanied by vertical subsidence, anomalous distribution of gravity and apparent resistivity, and the existence at its southern boundary on the St. Lawrence of a large Palaeozoic meteorite impact crater and localized seismicity. The area lies within the Grenville Province of the Precambrian Shield and contains mainly quartzofeldspathic gneisses which were folded, intruded and stabilized 800 to 1,100 million years ago. Anorthosite massifs occur in several places. A major thrust fault lying beneath the St. Lawrence, Logan's Line, marks the southern boundary of the Grenville and separates the Precambrian Shield from the Appalachian system to the south. A second line of weakness is implied by faulting along the Saguenay River between Lac Saint-Jean and the St. Lawrence, though its depth and extent are not well known. KUMARAPELI and SAULL (1966) have referred to the structure as the Saguenay graben.

Highly elevated parts of the Laurentide region are indicated in Fig. 1. DUNBAR and GARLAND (1975) have determined a positive free air gravity anomaly pattern

Contribution from the Earth Physics Branch No. 701.

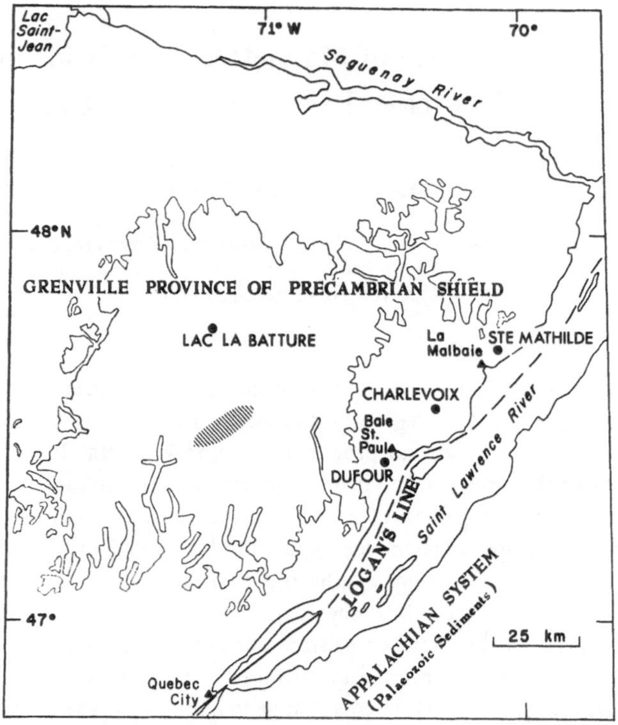

Fig. 1. Location of magnetotelluric stations in the St. Lawrence-Saguenay region. Areas inside the contours have elevations over 2,000 feet (610 m). The hatching indicates the approximate location of maximum subsidence.

which correlates closely with the topography implying that the elevated region represents an uncompensated load. Levelling data have been analysed by FROST and LILLY (1966) and VANICEK and HAMILTON (1972) and clear evidence of downward vertical crustal movement has been demonstrated. The downward movement again correlates closely with the free air gravity and elevation contours, the maximum rate of subsidence being about 45 cm per century relative to Quebec City. This behaviour is not typical of the Grenville Province nor indeed of the Canadian Shield which is undergoing a general uplift owing to post-glacial rebound. The stress differences created by the topographical load are not so great as to exceed the strength of the crust and in a normal region the load should be supported. The fact that it is not suggests that the downward motion may be a consequence of major crustal weaknesses associated with Logan's Line, the Saguenay graben and faulting in the western side of the triangle. Dunbar and Garland have suggested that these peripheral faults may effectively decouple the region from its surroundings.

The seismicity of the region has been documented by SMITH (1962, 1966) whose data compilations for eastern Canada are shown in Fig. 2. These indicate a well-confined zone of activity on the St. Lawrence River between Quebec City and the mouth of the Saguenay River. More precise data acquired by the Earth Physics Branch since 1960 have confirmed the existence of this restricted active zone. In

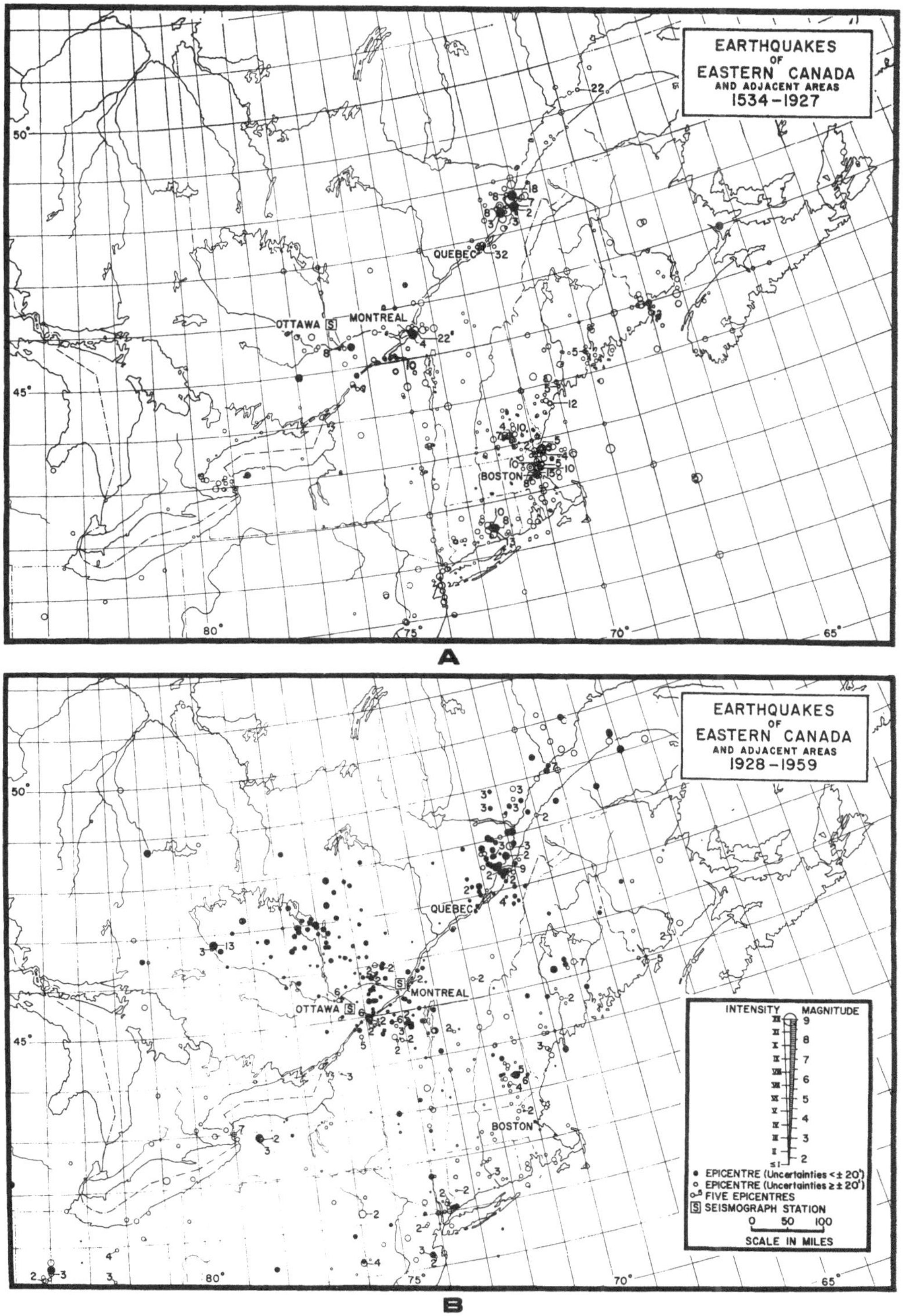

Fig. 2. Epicentral map for eastern Canada (SMITH, 1962, 1966).
A, Historical data. B, Instrumental data.

addition, microearthquake experiments in the locality by LEBLANC *et al.* (1973) and LEBLANC and BUCHBINDER (1977) reveal tightly confined seismicity over a distance of about 70 km along the St. Lawrence and centred near the town of La Malbaie on the north shore. This zone lies about 100 km northeast of the maximum vertical movement but is contiguous with Logan's Line underlying the river. ROBERTSON (1968, 1975) and RONDOT (1968) have identified a major meteorite impact crater of Palaeozoic age in Charlevoix County between the towns of Baie St. Paul and La Malbaie and corresponding very closely with the region of seismicity. The Charlevoix crater has a radius of 27 km and a depth of roughly 10 km. It extends beneath the St. Lawrence and undoubtedly intersects Logan's Line. Microearthquake activity appears to originate in the Precambrian rock beneath the river. Focal depths are about 10 km.

KURTZ and GARLAND (1976) and HONKURA *et al.* (1977) have examined magnetotelluric data from a few stations within 100 km of the Charlevoix crater and have found a conductive structure in the crust commencing about 40 km to the west. The extent of this anomalous conductor is not known; it lies near and probably includes the centre of vertical crustal movement and its eastern boundary appears to strike a little to the east of north.

DUNBAR and GARLAND (1975) suggest that the measured Laurentide subsidence is caused by outflow in the underlying low viscosity layer beneath the lithosphere. It is clear that this movement belongs to recent times and is not associated with the ancient orogenies which define the tectonics of the region. Its origin may lie in the glacial forebulge effect acting within a zone of partially decoupled crust as Dunbar and Garland have pointed out. The seismicity at La Malbaie is possibly related to the stresses associated with the anomalous vertical movement. It is perhaps not surprising that the earthquake energy is released mainly within and near the weakened and shattered crust of the Charlevoix crater and its intersection with a major tectonic boundary.

2. Magnetotelluric Experiments in and near the Charlevoix Region

The well-documented history of localized seismicity near La Malbaie suggests that this area should be more suitable than any other in eastern Canada for studying earthquake precursors and assessing various techniques for earthquake prediction. According to K. Whitham (personal communication, 1977) the estimated return periods are 3 1/2 years for earthquakes with magnitudes $M \geqq 4$, 14 years for $M \geqq 5$, 60 years for $M \geqq 6$ and about 230 years for $M \geqq 7$. Since 1900 the most prominent events have been those of 1925 ($M=7$), 1931 ($M=5.4$) and 1935 ($M=5.8$). The comparatively low level of activity in the last few decades suggests that the region is accumulating stress. Accordingly the Earth Physics Branch recently began a long-term study in and near Charlevoix County to measure geophysical parameters which might be sensitive to crustal deformation. Apparent resistivity was one of the parameters selected because of the encouraging results achieved in the Soviet

Union (BARSUKOV, 1972) and later in the U.S.A. (MAZZELLA and MORRISON, 1974). The process by which electrical resistivity changes occur in the earth's crust is not completely understood. However, laboratory work has shown that large stresses on rocks cause an inelastic increase in volume with a resultant opening and extension of micro-cracks. It has also been noted that in situ measurements of crustal rocks by deep electrical sounding techniques indicate resistivities several orders of magnitude less than values obtained in the laboratory on dry rock specimens. Since temperature is usually not high enough at crustal depths to account for this discrepancy, it is likely that fluid in the micro-cracks is contributing to the conductivity of the rocks. BRACE and ORANGE (1966, 1968a, b) measured the resistivity of water-saturated granite and other crystalline rocks under stress and observed decreases in resistivity as much as one order of magnitude prior to their fracture. This was attributed to fluid filling the newly formed micro-cracks.

HONKURA *et al.* (1976) have described a preliminary experiment in which continuous magnetotelluric (MT) measurements were made at a station inside the Charlevoix crater. Over a 12-month period a prominent and apparently real time dependence in the measured impedance tensor was detected, but no earthquakes exceeding magnitude 3 had occurred in the area and the cause of the impedance variations was not clear. A possible association with salinity and temperature variations in the St. Lawrence River was suggested, but more data were needed before one could hope to distinguish between seasonal fluctuations of this kind and the more random changes in resistivity which might be diagnostic of on-going deformation in the subsurface rock.

The Charlevoix station has therefore remained in operation and a second MT system was established in October 1975 at Lac la Batture, 70 km to the northwest. Its location in the Laurentide highlands is remote from the St. Lawrence River and presumably well beyond its influence. Data have also been available since December 1975 at two locations, Manic and Outardes, near a large new dam on the Manicouagan River about 265 km to the northeast (see Fig. 3). Some simultaneous

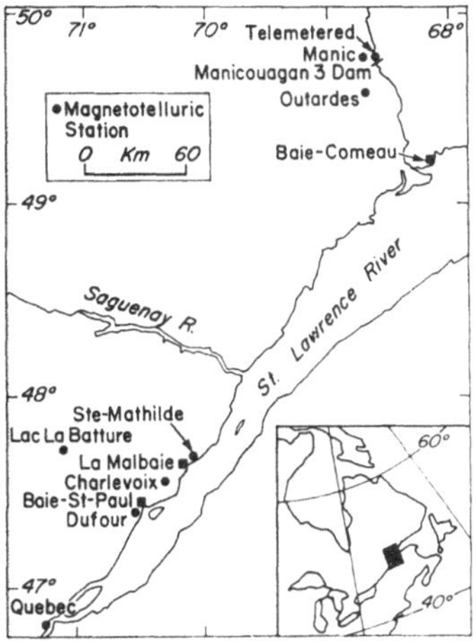

Fig. 3. Location map of seven magnetotelluric stations.

recordings from these four stations are now available for comparison. Data sequences have unfortunately suffered prolonged interruptions at all locations because of instrument failures, vandalism and the usual problems associated with unattended recording in remote areas.

The Manic and Outardes stations were established to observe changes in resistivity that may be associated with loading and seismicity caused by the filling of the Manicouagan 3 reservoir. The Manic station is about 6 km to the west of the centre of the earthquake zone associated with the dam; Outardes, which serves as a reference station, is 30 km to the south. In September 1976, the telemetered station was established on an island in the reservoir near the centre of induced seismicity. However, the activity has died away and significant changes in magnetotelluric parameters has not been demonstrated at these stations.

3. Instrumentation

Each magnetotelluric station consists of a magnetometer, a telluric system and a cassette recorder all powered by batteries. The magnetometer (TRIGG *et al.*, 1970) is a three-component fluxgate which measures north-south, east-west and vertical components of the earth's geomagnetic field variations. The telluric system (TRIGG, 1972) measures potential differences between north-south and east-west electrodes. Electrode separation is between 200 to 400 m and the input impedance of the system is greater than 10^9 ohms. The telluric channels are filtered with a pass band of 120–30,000 sec and the magnetic channels are low-pass filtered with the -3 dB roll-off occurring at 120 sec. The five channels are sampled once a minute and recorded digitally on a DATEL cassette recorder.

4. Data Analyses

The data are played back from the cassettes, edited and plotted. The records are then visually inspected and as many 12-hr sections of good data are selected as possible. Local daytime data usually yield the most stable results especially if quiet and very disturbed data are omitted. Smoothed spectral estimates are normally calculated from about 10 twelve-hour sections per cassette. Any estimates that have power less than a specified minimum are rejected. In addition, any estimates which give a multiple coherency between computed and measured electric fields less than 0.9 are rejected (REDDY *et al.*, 1976). The remaining estimates are averaged into frequency bands from which various parameters are calculated.

The parameters discussed in this paper are the impedance tensor elements and polarization characteristics of the telluric fields. The impedances, Z_{xx}, Z_{xy}, Z_{yx} and Z_{yy}, are computed from an assumed linear relationship between telluric and horizontal magnetic components given by

$$E_x = Z_{xx} H_x + Z_{xy} H_y$$
$$E_y = Z_{yx} H_x + Z_{yy} H_y \, .$$

E_x and E_y are the telluric components and H_x and H_y the horizontal magnetic components with x and y the measuring axes in geomagnetic north and east respectively. The impedance elements may be computed from the spectral estimates (see SIMS *et al.*, 1971) and rotated into the major axis of anisotropy. Similarly, the spectral estimates may be used to calculate the polarization characteristics of the telluric and magnetic fields (FOWLER *et al.*, 1967). The parameters from each 12-hr section have been averaged. Some of these averaged parameters along with error bars of two standard deviations are plotted in Figs. 4, 6, 7, and 8. With the procedures used here it requires about 17 days of continuous recording to produce sufficient data for reliable computation of the impedance tensor. Multiple coherency values for the final averages were greater than 0.96.

Any measurement of the magnetic or telluric field is likely to contain some noise. Because of the method used to compute the impedance tensor elements (see Eq. 15, SIMS *et al.*, 1971) random noise on the telluric channels will not affect the impedance estimates but random noise on the magnetic signal will bias these estimates downwards. However, analysis of simultaneous data from Charlevoix and Lac la Batture (70 km distant) indicate that power ratios of the magnetic components do not vary by more than 9%. REDDY *et al.* (1976) demonstrate how the amount of downward bias may be estimated from the multiple coherence. Because of the stringent criteria we have used for selection of good data the impedances shown in this paper could not be biased downward by more than 5 to 10% using their assumptions.

Bias of spectral estimates should not therefore lead to serious errors in the computed impedance values. If the noise is continuous then the bias effect is not included in the error bars associated with the data plotted in Figs. 4, 6, 7, and 8. If the noise is intermittent then the bias effect will contribute to the error bars.

5. Results

The coordinates and operating times of the stations shown in Fig. 3 are given in Table 1. Only the first five listed have provided sufficient data for analysis at the present time. These all display a high degree of anisotropy. The electric fields are

Table 1. MT stations.

	Latitude	Longitude	Period of operation	Skew	
				5 min	10 min
Charlevoix	47°33.0′	70°19.6′	Oct. 9, 1974–present	0.40	0.40
Lac la Batture	47°65.6′	71°13.0′	Oct. 15, 1975–present	0.30	0.35
Manic	49°50.9′	68°42.8′	Dec. 17, 1975–Aug. 31, 1977	0.20	0.18
Outardes	49°34.5′	68°41.8′	Dec. 19, 1975–Aug. 31, 1977	0.60	0.60
Telemetered	49°50.3′	68°37.2′	Oct. 16, 1976–July 31, 1977	0.15	0.10
Ste-Mathilde	47°41.6′	70°05.7′	June 15, 1977–present	—	—
Dfurou	47°23.8′	70°32.3′	Aug. 29, 1977–present	—	—

highly polarized in a fixed direction. Due to the complex geology found at all station sites, it is not surprising that the impedance tensors show high skew and strong three-dimensional properties at Charlevoix, Lac la Batture and Outardes.

The bottom two traces in Fig. 4 show the time-dependence of impedance measured along the major axis of anisotropy at the Charlevoix station. The actual measured impedances are not plotted here, but instead the ratio of each individual value to the arithmetic mean of all values. The variations as illustrated in Fig. 4 and subsequent figures are therefore normalized to the mean value of the measured earth impedance at the station and do not depend on whether the impedance itself is large or small. Note the unfortunate data gap from November 1975 to June 1976.

The impedance at Charlevoix has changed 50% or more over the past 3 years

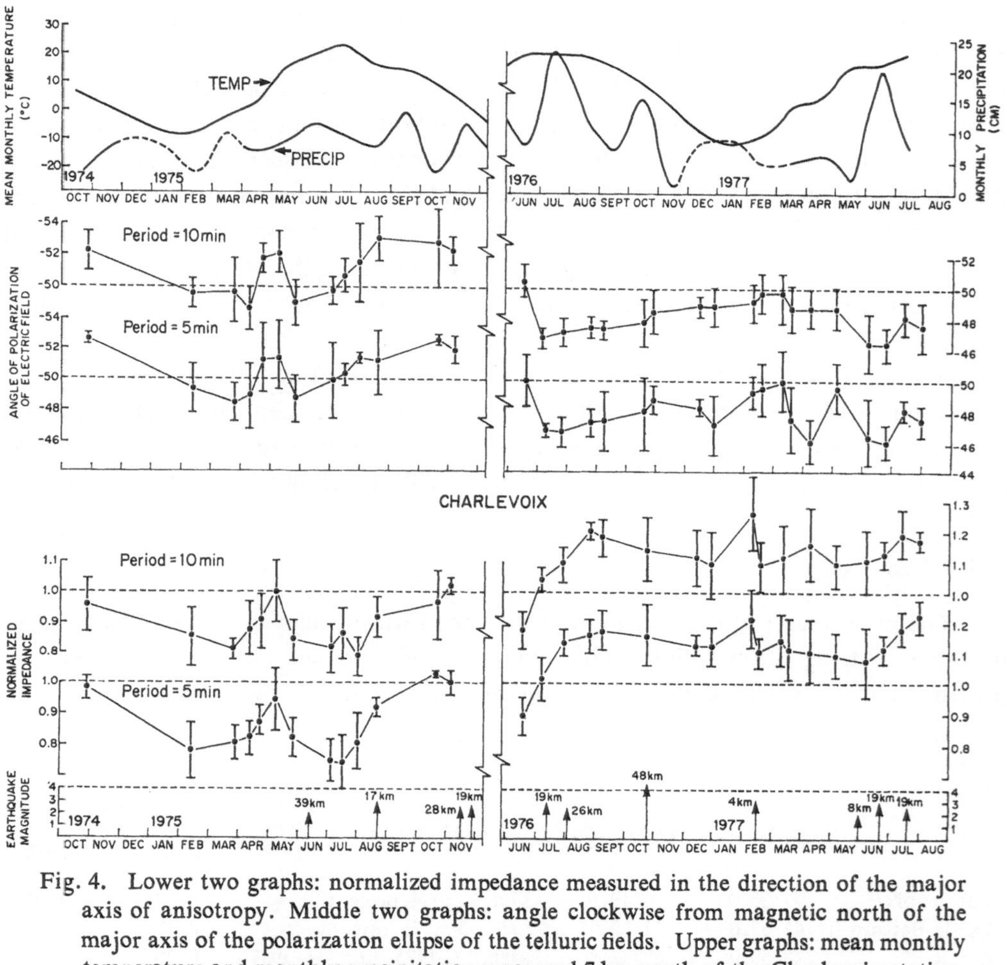

Fig. 4. Lower two graphs: normalized impedance measured in the direction of the major axis of anisotropy. Middle two graphs: angle clockwise from magnetic north of the major axis of the polarization ellipse of the telluric fields. Upper graphs: mean monthly temperature and monthly precipitation measured 7 km south of the Charlevoix station. The dashed line part indicates that precipitation was probably in the form of snow. Error bars are two standard deviations in length. Earthquakes of magnitude (m_N or M_L) greater than 2 that have been detected within 50 km of the Charlevoix station are shown as well as their approximate distance from the station.

with changes of up to 30% occurring over a few months for the 5-min period data. The changes become less pronounced at longer periods. Earthquakes of magnitude greater than 2 that have been reported by the Canadian Seismograph Network within a radius of 50 km of the Charlevoix station are indicated by vertical arrows across the bottom of Fig. 4. The only earthquake greater than magnitude 4 occurred approximately 48 km to the north-east on October 23, 1976. At this distance it is unlikely that the Charlevoix station would detect any premonitory changes in earth resistivity. Therefore it is doubtful that the 30% increase in impedance over the 4 months preceding the earthquake can be associated with it.

The impedance shows a step which appears to have coincided with the February 14, 1977 earthquake of magnitude 3.2. However, the scatter bars overlap appreciably and there is probably little justification for considering the event to be significant at this time.

The middle two traces in Fig. 4 show the polarization angle of the electric field for 5- and 10-min period variations. The measured electric field is over 97% linearly polarized at the Charlevoix station. If changes in resistivity of the earth's crust are occurring, the angle of polarization may be a sensitive indicator since the current flow in the earth will readjust in amplitude and direction. The angle does indeed change in a fashon similar to the impedance but again the measured changes do not appear to correlate with the seismicity.

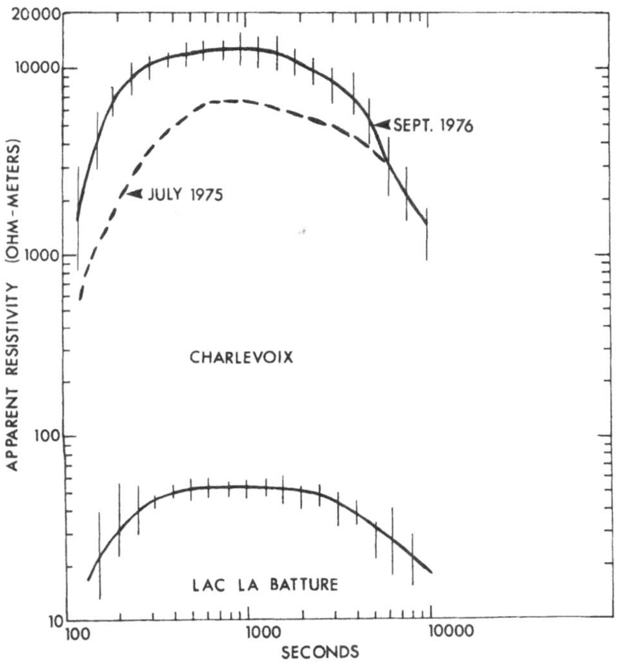

Fig. 5. Apparent resistivity curves for Charlevoix and Lac la Batture calculated along the major axis of anisotropy. Charlevoix has been computed for a time of maximum impedance and for a time of minimum impedance.

It is interesting to note the effect of temporal variations of impedance on the apparent resistivity curves which are usually displayed and interpreted in an MT analysis. Figure 5 shows the apparent resistivity values calculated at Charlevoix at a time of maximum impedance in September 1976 and at a time of minimum in July, 1975. The maximum peaks at 13,000 Ω m at a period of approximately 1,000 sec while the lower curve peaks at 7,500 Ω m representing a change of about 70%. Also plotted is the apparent resistivity curve in the major axis of anisotropy for Lac la Batture. This demonstrates the dramatic difference in earth resistivity between the two stations. Lac la Batture is located over the conductive structure in the crust referred to earlier.

What causes the temporal impedance variations? There are no apparent correlations between daily rainfall or temperature fluctuations and measured impedance at Charlevoix. Nor does there appear to be any obvious relation between monthly averages of precipitation or temperature as can be seen in Fig. 4.

Are these impedance changes a widespread phenomenon? To help answer this question and to check the long-term stability of our instruments, comparisons were made with measured impedances at Lac la Batture and in the Manicouagan region at the telemetered station and the stations of Manic and Outardes (Fig. 3 and Table 1). Figure 6 shows the normalized impedance at these four stations for 5-min period variations. While only Charlevoix appears to display significant time variations, a gradual increase of impedance over the last 20 months may also have occurred at Outardes. The data plotted in Fig. 7 suggest that the same conclusion will hold for 10-min period variations.

It is important to compare between stations a statistical measure of the time dependence in impedance (Z_p) along the principal axis of anisotropy. Each Z_p value

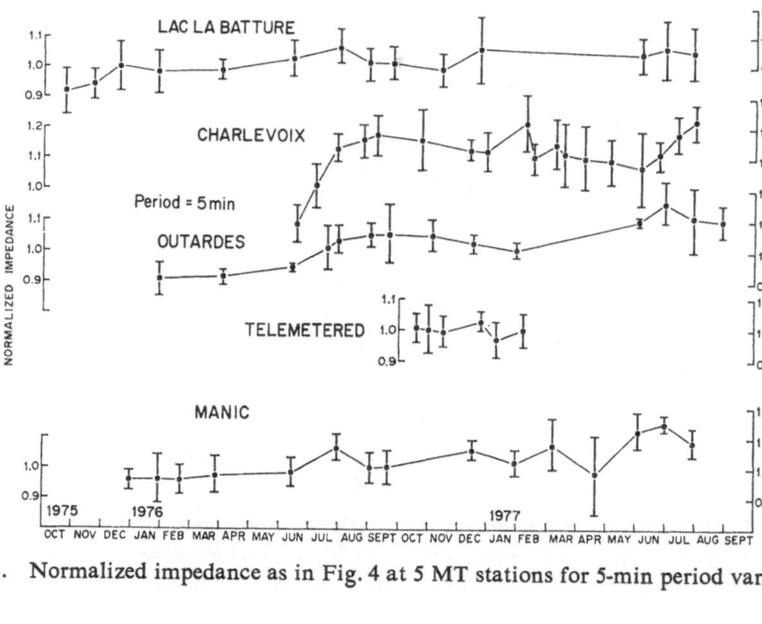

Fig. 6. Normalized impedance as in Fig. 4 at 5 MT stations for 5-min period variations.

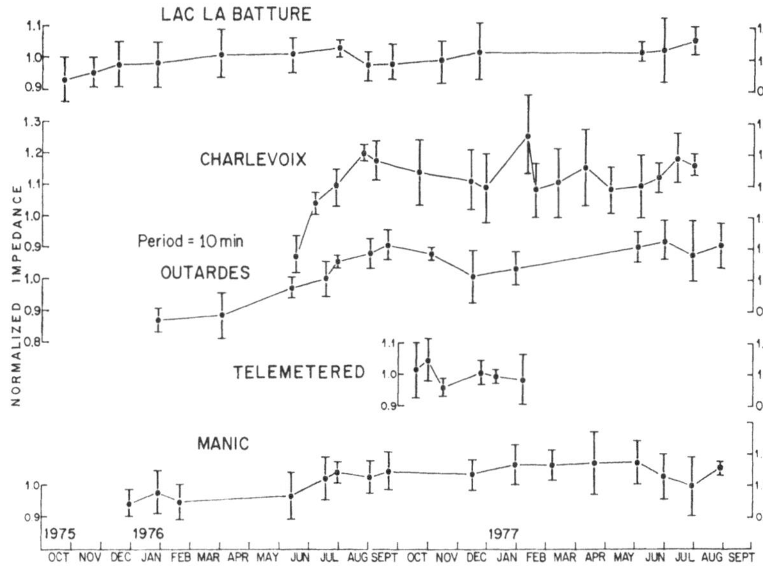

Fig. 7. Normalized impedances for 10-min period variations.

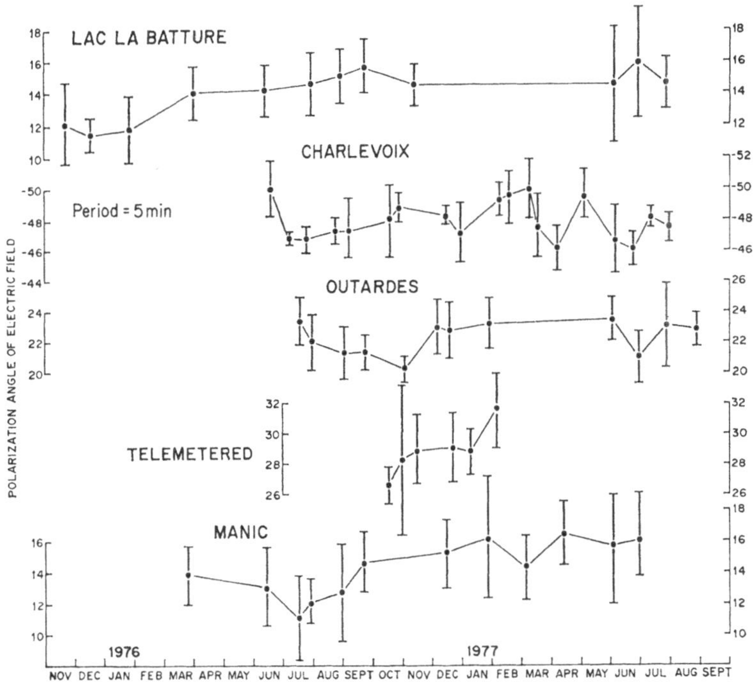

Fig. 8. Angle clockwise from magnetic north of the major axis of polarization ellipse of the telluric field at 5 MT stations for 5-min period variations.

computed from power spectrum estimates of the E and H fields has associated with it a standard error σ. The error bars shown on the plots in Figs. 4, 6, 7, and 8 are each 2σ in length. To indicate whether or not the measured time variations are significant we compute the rms deviation ϕ of the individual values of impedance about the mean where

$$\phi=\left\{\sum_{i=1}^{N}\frac{(Z_{p_i}-\bar{Z}_p)^2}{N}\right\}^{1/2}.$$

Here N is the total number of impedance estimates in the time sequence, Z_{p_i} is an individual measured value, and $\bar{Z}_p$ is the arithmetic mean. Values of $\bar{Z}_p$, N, ϕ, $\bar{\sigma}$ and σ_{max} are illustrated in Table 2 for all five stations for both 5- and 10-min periods. $\bar{\sigma}$ is the mean of all N standard deviation estimates and σ_{max} is the largest. In these calculations Z_p is expressed in its measured units (mV/km nT) and is not represented as a normalized ratio. The ratio $\phi/\bar{\sigma}$ is a measure of the variability of Z_p, the inference being that the measured values are time-dependent if $\phi/\bar{\sigma}>1$.

For the 5-min period data Table 2A shows that ϕ is over twice as large as $\bar{\sigma}$ at the Charlevoix station and appreciably larger than the largest value of σ in the time sequence. The measured time dependence at Charlevoix cannot therefore be explained by errors in experimentation or in power spectrum estimates and must be

Table 2. Variation parameters for impedance (Z_p mV/km nT) along the major axis of anisotropy.

A. 5-min period

	Charlevoix	Lac la Batture	Manic	Outardes	Telemetered
$\bar{Z}_p$	11.31	0.893	3.93	3.53	4.64
N	32	14	15	15	6
ϕ	1.66	0.040	0.24	0.23	0.083
$\bar{\sigma}$	0.74	0.059	0.21	0.16	0.24
σ_{max}	1.29	0.098	0.50	0.37	0.36
$\phi/\bar{\sigma}$	2.24	0.67	1.12	1.54	0.34
$\phi'/\bar{\sigma}$	1.22	0.42	0.68	0.83	0.33
B	14.3	—	9.2	12.4	—
R	0.84	—	0.79	0.84	—

B. 10-min period

	Charlevoix	Lac la Batture	Manic	Outardes	Telemetered
$\bar{Z}_p$	8.95	0.651	2.78	2.57	3.38
N	31	14	17	15	6
ϕ	1.22	0.021	0.119	0.20	0.089
$\bar{\sigma}$	0.70	0.039	0.155	0.14	0.18
σ_{max}	1.30	0.064	0.27	0.27	0.30
$\phi/\bar{\sigma}$	1.75	0.53	0.76	1.39	0.48
$\phi'/\bar{\sigma}$	0.95	0.35	0.48	0.89	0.45
B	13.3	—	—	12.8	—
R	0.84	—	—	0.77	—

considered to be statistically significant. This situation does not apply at the other four stations. At Lac la Batture ϕ is smaller than $\bar{\sigma}$ and this is also the case at the telemetered station. At Outardes and Manic ϕ lies between $\bar{\sigma}$ and σ_{max} suggesting that time dependence is a possiblility though the effect is much less prominent than at Charlevoix. The parameters in Table 2A therefore clearly imply that Z_p at Charlevoix is subject to real variations with time and in this respect behaves differently from Z_p measured at the other stations. Quite similar results are evident in Table 2B for the 10-min period data, though the time dependence is less pronounced at Charlevoix for periods of 10 min and longer. Z_p is again marginally variable at Outardes, but not at Manic.

Least-squares straight lines have been fitted to the data at those stations for which $\phi/\bar{\sigma}>1$. In Table 2, B represents the slope of the linear trend (expressed as percentage change per year), R is the correlation coefficient which gives a measure of the degree to which Z_p and time are linearly related, and $\phi'/\bar{\sigma}$ is the variability ratio re-computed after removal of the linear trend from the data. $\phi'/\bar{\sigma}$ is greater than unity only at Charlevoix for the 5-min data. Thus while much of the time dependence at Charlevoix is caused by the trend increase of 14% per year, an appreciable part remains which must be quasi-periodic or non-linear. This result is confirmed by application of the χ^2 test to the residuals after removal of the trend.

Table 3. Variation parameters for polarization azimuth ($\alpha°$) of the electric field.

A. 5-min period

	Charlevoix	Lac la Batture	Manic	Outardes	Telemetered
$\bar{\alpha}$	−49.1	14.1	14.3	22.2	28.9
N	32	12	12	13	6
ϕ	1.80	1.42	1.70	1.11	1.50
$\bar{\sigma}$	1.41	2.02	2.48	1.64	2.39
σ_{max}	2.6	3.6	3.9	2.8	5.0
$\phi/\bar{\sigma}$	1.28	0.70	0.68	0.67	0.63
$\phi'/\bar{\sigma}$	0.99	0.46	0.44	0.64	0.32
B	2.7	—	—	—	—
R	0.63	—	—	—	—

B. 10-min period

	Charlevoix	Lac la Batture	Manic	Outardes	Telemetered
$\bar{\alpha}$	−49.4	16.4	15.9	22.1	30.8
N	32	12	12	13	6
ϕ	1.84	1.59	0.97	1.26	1.58
$\bar{\sigma}$	1.30	2.05	2.74	2.11	2.72
σ_{max}	2.8	2.7	4.9	4.8	3.4
$\phi/\bar{\sigma}$	1.41	0.77	0.35	0.60	0.58
$\phi'/\bar{\sigma}$	1.07	0.36	0.25	0.58	0.22
B	2.8	—	—	—	—
R	0.65	—	—	—	—

α is positive east of north.

Quadratic or higher order polynomial terms are required for an adequate fit to the 5-min data at Charlevoix; at Outardes and Manic there is no evidence for variability apart from the measured trend.

Figure 8 shows the azimuth α of the direction of telluric polarization for 5-min period variations at the five stations. All plots suggest changes with time ranging over an interval of about 4°, and a significance test has been applied as before. Table 3 gives the rms deviation (ϕ) for values of α about its mean along with other variation parameters defined for Table 2. The ratio $\phi/\bar{\sigma}$ in Table 3 indicates that the polarization angle is marginally variable at Charlevoix at the 5- and 10-min periods. The χ^2 test confirms that for both periods this variability can be adequately accounted for by a linear trend. Thus time dependence in the impedance tensor at Charlevoix is mainly reflected by changes in Z_p rather than α.

6. Conclusions

It seems clear that the impedance computed from MT fields in Charlevoix is subject to real time dependence and that the changes are a local rather than a widespread effect. The Charlevoix station also differs from the other recording sites in that it lies close to the St. Lawrence River (about 8 km from the north shore), it lies within a large meteorite crater and it lies near the centre of a zone of local seismicity. HONKURA *et al.* (1976) suggested that the observed changes might be controlled by changes in salinity of the St. Lawrence River caused by seasonal temperature variations, by rainfall and by melting snow and ice during the spring runoff. However, the longer sequence of data presented here does not display effects that can be clearly identified as seasonal trend and the rapid increase and decrease in impedance that occurred in April–June 1975 has not repeated in the following years. Monthly precipitation, particularly rainfall, does not appear to be a factor (Fig. 6) but more data are required to establish whether or not a meaningful correlation exists with the impedance. Though the results obtained so far suggest that these parameters are uncorrelated, both rainfall and impedance display somewhat similar periodicities and the proximity of the St. Lawrence (here about 20 km wide) might cause the telluric field to be more sensitive to rainfall at Charlevoix than at the inland stations. On the other hand, the average impedance seems to have increased by over 40% between 1975 and 1977 and this trend, which forms the more important part of the time dependence, cannot be associated with local rainfall or precipitation.

Over the 3-year recording interval Charlevoix County has experienced fairly continuous low-level seismicity indicative of on-going local crustal instability. The non-linear part of the variable impedance determined from the magnetotelluric data at Charlevoix suggests that electrical conductivity at upper crustal depths may be changing in response to this instability. This conclusion is supported by the absence of both seismicity and comparable impedance variations at Lac la Batture and the other recording sites but other factors may be important. The possibility that the telluric field at Charlevoix is influenced by seasonally changing conditions in the St.

Lawrence River is as yet unresolved. Further the short-period impedance changes, while significant in a statistical sense, may owe their existence to interaction between gradual changes in position or configuration of the source fields and the complex three-dimensional structure that is characteristic of the Grenville Province. The source fields may be expected to be similar at the various recording sites at any given time, but the crustal structure is complex and may differ substantially at each station. The interaction between seasonal changes in the aspect and polarization of the sources and structure might create a time dependence which is ficticious in the sense that it is unrelated to changing electrical parameters in the crust. This problem requires further examination. In spite of these limitations it seems unlikely that the long-term increasing trend in impedance (Fig. 6) at Charlevoix can be explained by either superficial surface features or by variable source fields. The evidence supporting changing conditions at depth in the crust is therefore stronger than formerly (HONKURA *et al.*, 1976).

With the data presented here it has not been possible to demonstrate a direct correspondence between impedance change and a specific earthquake occurrence. As well, there have been no observed significant changes in seismic P velocities detected by means of calibration explosions in the Charlevoix region between 1974 and 1976 (BUCHBINDER and KEITH, 1977). Results from precise gravity network studies as well as measurements of strain and tilt at the Charlevoix station have not displayed any conclusive precursory phenomena (LAMBERT *et al.*, 1977). However, there have been occasional and perhaps completely accidental correlations between tilt and strain readings and small local earthquakes (J.J. Labrecque, personal communication, 1977). The absence of premonitory effects is likely due to the fact that the largest earthquake to occur in the immediate vicinity of the station during these experiments had a magnitude of only 3.2. Earthquakes of such low magnitude probably do not produce premonitory effects large enough to be detected by the MT observations we have made.

The station at Lac la Batture has shown that changes in impedance do not extend inland from the St. Lawrence River and that stable earth parameters can be obtained by the magnetotelluric method. However, we now require additional stations situated in a geographical and geological setting more similar to Charlevoix since Lac la Batture has apparent resistivity values which are two orders of magnitude less. Also the standard error in the impedance estimates at these stations is typically about 7 or 8% indicating that a closer network of stations is needed for evaluation of source field effects. For these reasons two additional stations, Ste-Mathilde and Dufour (Fig. 1) were established in June 1977. These are within 25 km of Charlevoix and situated approximately the same distance from the St. Lawrence River. They will enable a better estimate to be made of the region over which resistivity changes are occurring.

Hydro-Québec generously provided manpower, machines and lodging in the Manicouagan area. In particular Mr. Pierre Giroux and Mr. Pierre Tassé provided assistance and advice. Flavius Bérubé and Denis Dubois regularly serviced our stations even during the harsh Canadian winters, for which we are very grateful.

We thank the Québec Ministère du Tourisme, de la Chasse et de la Pêche for kind permission to operate the Lac la Batture station in Laurentides Park.

Dr. P.A. Camfield, Mr. J.M. DeLaurier, Mr. D.F. Trigg, Mr. R. Groulx, Mr. E. Berndt, Mr. J.J. Labrecque, Miss M.C. Fortier, and Dr. K. Whitham gave valuable assistance during preparation, installation and servicing of the instruments. We thank Dr. P.H. Serson for his guidance and encouragement throughout the experiment.

REFERENCES

BARSUKOV, O.M., Variations of electric resistivity of mountain rocks connected with tectonic causes, *Tectonophysics*, **14**, 273–277, 1972.

BRACE, W.F. and A.S. ORANGE, Electrical resistivity changes in saturated rock under stress, *Science*, **153**, 1525–1526, 1966.

BRACE, W.F. and A.S. ORANGE, Electrical resistivity changes in saturated rocks during fracture and frictional sliding, *J. Geophys. Res.*, **73**, 1433–1445, 1968a.

BRACE, W.F. and A.S. ORANGE, Further studies of the effects of pressure on electrical resistivity of rocks, *J. Geophys. Res.*, **73**, 5407–5420, 1968b.

BUCHBINDER, G. and C. KEITH, Search for changes in velocity in the La Malbaie region of Quebec, *EOS, Trans. Am. Geophys. Union*, **58**, 433, 1977 (Abstr.).

DUNBAR, W.S. and G.D. GARLAND, Crustal loads and vertical movements near Lake St. John, Quebec, *Can. J. Earth Sci.*, **12**, 711–720, 1975.

FOWLER, R.A., B.J. KOTICK, and R.D. ELLIOTT, Polarization analysis of natural and artificially induced geomagnetic micropulsations, *J. Geophys. Res.*, **72**, 2871–2883, 1967.

FROST, N.H. and J.E. LILLY, Crustal movement in the Lake St. John area, Quebec, *Can. Surv.*, **20**, 292–299, 1966.

HONKURA, Y., R.D. KURTZ, and E.R. NIBLETT, Geomagnetic depth sounding and magnetotelluric results from a seismically active region northeast of Quebec City, *Can. J. Earth Sci.*, **14**, 256–267, 1977.

HONKURA, Y., E.R. NIBLETT, and R.D. KURTZ, Changes in magnetic and telluric fields in a seismically active region of eastern Canada: Preliminary results of earthquake prediction studies, *Tectonophysics*, **34**, 219–230, 1976.

KUMARAPELI, P.S. and V.A. SAVLL, The St. Lawrence Valley System: A North American equivalent of the East African Rift Valley System, *Can. J. Earth Sci.*, **3**, 639–658, 1966.

KURTZ, R.D. and G.D. GARLAND, Magnetotelluric measurments in eastern Canada, *Geophys. J. R. Astr. Soc.*, **45**, 321–347, 1976.

LAMBERT, A., J. LIARD, and H. DRAGERT, Canadian precise gravity networks for crustal movement studies; an instrument evaluation, *Tectonophysics*, **52**, 87–96, 1979.

LEBLANC, G. and G.G.R. BUCHBINDER, Second microearthquake survey of the St. Lawrence Valley near La Malbaie, Quebec, *Can. J. Earth Sci.*, **14**, 2778–2789, 1977.

LEBLANC, G., A.E. STEVENS, R.J. WETMILLER, and R. DU BERGER, A micro-earthquake study of the St. Lawrence Valley near La Malbaie, Quebec, *Can. J. Earth Sci.*, **10**, 42–53, 1973.

MAZZELLA, A. and H.F. MORRISON, Electrical resistivity variations associated with earthquakes on the San Andreas Fault, *Science*, **185**, 855–857, 1974.

REDDY, I.K., R.J. PHILLIPS, J.H. WHITCOMB, D.M. COLE, and R.A. TAYLOR, Monitoring of time dependent electrical resistivity by magnetotellurics, *J. Geomag. Geoelectr.*, **28**, 165–178, 1976.

ROBERTSON, P.B., La Malbaie structure, Quebec—A Palaeozoic meteorite impact site, *Meteoritics*, **4**, 1–24, 1968.

ROBERTSON, P.B., Zones of shock metamorphism at the Charlevoix impact structure, Quebec, *Bull. Geol. Soc. A*, **86**, 1630–1638, 1975.

RONDOT, J., Nouvel impact meteoritique fossile? La structure semicirculaire de Charlevoix. *Can. J. Earth Sci.*, **5**, 1305–1317, 1968.

SIMS, W.E., F.X. BOSTICK, and H.W. SMITH, The estimation of magnetotelluric impedance tensor elements from measured data, *Geophysics*, **36**, 938-942, 1971.

SMITH, W.E.T., Earthquakes of eastern Canada and adjacent areas 1534-1927, *Publ. Dom. Obs., Ottawa*, **26**, 271-301, 1962.

SMITH, W.E.T., Earthquakes of eastern Canada and adjacent areas 1928-1959, *Publ. Dom. Obs., Ottawa*, **32**, 87-121, 1966.

TRIGG, D.G., An amplifier and filter system for telluric signals, *Publ. Earth Phys. Branch, Ottawa*, **44**, 1-5, 1972.

TRIGG, D.G., P.H. SERSON, and P.A. CAMFIELD, A solid state electrical recording magnetometer. *Publ. Earth Phys. Branch, Ottawa*, **41**, 67-80, 1970.

VANICEK, P. and A.C. HAMILTON, Further analysis of vertical crustal movement observations in the Lac St. Jean area, Quebec, *Can. J. Earth Sci.*, **9**, 1139-1147, 1972.

Piezomagnetic Response with Depth, Related to Tectonomagnetism as an Earthquake Precursor

R.S. CARMICHAEL

Department of Geology, University of Iowa, Iowa City, Iowa 52242, U.S.A.

(Received September 16, 1977)

Piezomagnetic field variations can result from tectonic stress changes in the focal zone of an impending earthquake. Interpretation of such observed tectonomagnetic effects requires modeling of the stress response of magnetic properties of the lithosphere. Calculations have been made to estimate the here piezomagnetic effect as a function of depth, by considering the effect of stress on the magnetization of magnetite (Fe_3O_4) with increasing temperature and hydrostatic pressure. The responsiveness of magnetization to seismotectonic stress is gauged by an appropriate balancing of magnetocrystalline and magnetoelastic anisotropy energies. The calculations indicate that magnetite becomes progressively more responsive to stress at depth increases. The rate of change depends on the local geothermal gradient. The upper 15 km of the lithosphere is likely to be the most important in yielding observable piezomagnetic field effects. Such shallow-focus earthquakes are expected to be best for monitoring tectonomagnetic anomalies for earthquake prediction.

1. Introduction

For the task of earthquake prediction, the phenomenon of piezomagnetism—or tectonomagnetism, for relating observed surface fields to subsurface stress and geologic tectonism—has much potential value (RIKITAKE, 1968; NAGATA, 1972). Tectonic stresses acting in a focal zone region of the lithosphere cause transitory changes in the magnetic properties of rocks and thus of their observed magnetic field. An objective is reliable quantitative interpretation of these precursory piezomagnetic field changes in terms of the subsurface stress buildup. For this, one needs understanding of the response of rocks to stress, as a function of depth in the crust and upper mantle. Both temperature and hydrostatic pressure increase with depth, and these conditions affect the magnetization of rocks and how they respond to directed (tectonic) stress.

Relating the observed tectonomagnetic field anomalies to a focal-zone model requires integration of work in field monitoring of piezomagnetism in seismically-active regions (e.g. ISPIR and UYAR, 1971), laboratory study of piezomagnetic effects, and theoretical analysis of expected piezomagnetic effects (YUKUTAKE and TACHINAKA, 1967).

This is a theoretical calculation of the expected relative stress response of rocks with depth, to aid in refinement of modeling observed field changes for earthquake prediction. This paper was prepared for the IAGA "Symposium on Tectonomagnetics". A related article was in press (CARMICHAEL, 1977) by the time the Symposium was convened. Thus this note is a summary of the results, particularly as related to the companion papers here and interests of the Symposium.

2. Analysis

A calculation can be made to estimate the piezomagnetic effect as a function of depth. One approach is to consider the responsiveness to stress of ferrimagnetic magnetite, as conditions of temperature and hydrostatic pressure change with depth. The titanomagnetite series (magnetite-ulvospinel) is of particular importance because of its common occurrence in lithospheric rocks and dominance in determining their magnetic properties. The series is

$$(1-x)Fe_3O_4 \cdot xFe_2TiO_4$$

where 'x' represents the mole fraction of ulvospinel.

The relative response of magnetization to directed tectonic stress can be indicated (CARMICHAEL, 1968) by an appropriate balancing of magnetocrystalline anisotropy energy, which for the cubic anisotropy of magnetite is

$$E_K = f(K_1, K_2) = K_1(\alpha_1^2\alpha_2^2 + \alpha_2^2\alpha_3^2 + \alpha_3^2\alpha_1^2) + K_2(\alpha_1^2\alpha_2^2\alpha_3^2) + \cdots$$

where

K_1, K_2 = magnetocrystalline energy constants, ergs/cm^3
α_i = direction cosines of magnetization vector with respect to cubic axes

and of magnetoelastic anisotropy energy

$$E_{\lambda\sigma} = f(\lambda_{111}, \lambda_{100}) = -\frac{3}{2}\lambda_{100}\sigma(\alpha_1^2\gamma_1^2 + \alpha_2^2\gamma_2^2 + \alpha_3^2\gamma_3^2)$$
$$-3\lambda_{111}\sigma(\alpha_1\alpha_2\gamma_1\gamma_2 + \alpha_2\alpha_3\gamma_2\gamma_3 + \alpha_1\alpha_3\gamma_1\gamma_3)$$

where

λ_{111}, λ_{100} = magnetostriction constants, cm/cm
γ_i = direction cosines of stress vector with respect to cubic axes
σ = applied directed stress, dynes/cm^2; positive for tension.

The direction of domain alignment is determined by the anisotropy of the net energy, not just the relative magnitudes of different energy terms. The criterion used for the influence of directed stress on magnetization is removal or sufficient reduction of potential energy barriers between preferred 'easy' axes of spontaneous magnetization (J_s). This gives 'control' of magnetization orientation by rotation in single-domain grains, and rotation and domain wall motion in multidomain crystals. For sufficient stress, the domain arrangement and alignment is effectively influenced.

There will thus be stress-imposed changes in susceptibility and induced magnetization (NULMAN *et al.*, 1978), and in remanent magnetization.

The determining factor for the stress needed is the character of the energy barrier between adjacent axes of minimum magnetocrystalline energy. For cubic magnetite, these directions are the [111] axes. The intermediate energy barriers are the [110] directions. An applied stress gives rise to magnetoelastic energy anisotropy which favors deflection of domain magnetization to directions that depend on the stress orientation and nature of the material's magnetostriction.

The energy barrier will vary with depth in the earth because the constants K and λ vary with temperature and hydrostatic pressure. The intensity of spontaneous magnetization will decrease with depth because of the geothermal gradient, disappearing at the Curie temperature (T_c) isotherm. For magnetite at room pressure, T_c is 575°C; it is lower for increasing 'x' in titanomagnetite. Magnetic anomalies and tectonomagnetic field variations originate from the zone of the crust and upper mantle above the T_c depth.

For the theoretical calculation, we need quantitative data on temperature and pressure variation with depth, and the resulting effect on the magnetic constants (K_1, K_2, λ_{111}, λ_{100}, T_c). These data are presented elsewhere (CARMICHAEL, 1977), along with the appropriate calculations, and will be summarized here.

For typical geothermal gradients, the Curie-point depth for magnetite will occur at about 60 km below typical ancient shields and about 33 km below typical oceans. The 'effective' depth will be somewhat less, because the presence of titanium in

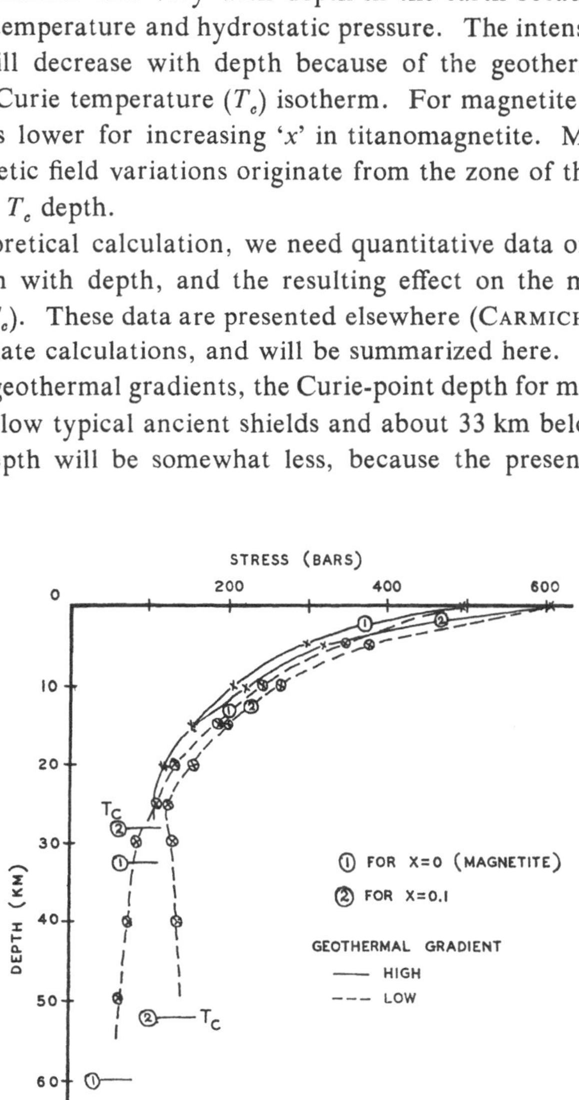

Fig. 1. Calculated stress response of magnetite with depth. 'Stress' is that required to deflect domain magnetization from [111] to [11$\bar{1}$] direction. From CARMICHAEL (1977).

titanomagnetite will lower the T_c, and thus result in a shallower T_c isotherm than for pure magnetite. Further, spontaneous magnetization decreases as the Curie temperature is approached. A summary of calculated and interpreted depths for studies of various continental regions is given by GREEN (1976).

The responsiveness to stress will be estimated by the relative directed stress needed for the magnetoelastic energy to overcome magnetocrystalline energy barriers, with the influence of temperature and hydrostatic pressure introduced.

Figure 1 shows the theoretical stress senstivity of magnetite ($x=0$) and titanomagnetite (with $x=0.1$). It is shown for regions of high ('typical ocean') and low ('typical shield') geothermal gradients. The response is represented (on the abscissa) as the directed stress required to give the equivalent piezomagnetic effect, i.e. control of magnetic domain pattern. The decreased stress needed with depth is a consequence of the magnetocrystalline energy barrier having been lowered because of the effects of temperature and hydrostatic pressure.

The trend of the stress sensitivity relation does not change significantly with titanomagnetite's titanium content varying from $x=0$ to 0.1.

Figure 1 shows the relative sensitivity to stress, for an arbitrary "base-level" stress—that required to effectively 'control' the orientation and arrangement of a magnetic domain pattern. This is not to imply that such a stress—e.g. about 500 bars for magnetite at room temperature—is a required 'threshold' to cause piezomagnetic effects, or that such a stress change is needed in a focal zone. Magnetization will change, in direction and intensity, for much lower applied stresses. Changes will be primarily reversible at the lower stresses. The figure (and stress values) indicates the relative ease of change, and not a threshold tectonic stress needed to produce any change at all. The curve could have been normalized to a 'surface stress' of 1.0, which could represent a stress of, say, 50, 100, or 497 bars. The 'ease' is reflected by the change in that initial stress. At depth, a lower stress is apparently needed to produce an equivalent piezomagnetic response.

In practice, the total differential tectonic stress expected in seismically-active regions may be up to several kilobars. The directed stress relieved in a focal zone by a moderate earthquake is variously estimated from a few tens of bars to perhaps 200 bars.

3. Discussion

The calculations indicate that magnetite becomes progressively more responsive to applied (directed) stress as depth increases. The rate of change depends on the local geothermal gradient. At a depth of 15 km only about a third as much tectonic stress is required to give the same tectonomagnetic effect as at surface conditions of temperature and pressure.

An estimate can now be made of the net piezomagnetic field effect to be observed at the earth's surface, for a stressed source (focal region) at different depths. For this, we must combine:

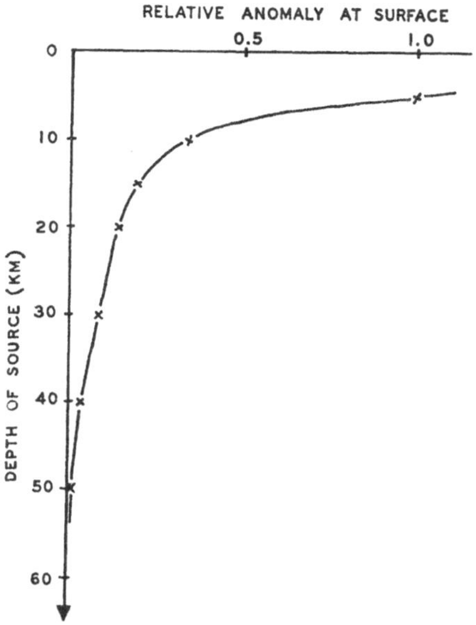

Fig. 2. Relative observed piezomagnetic anomaly expected for stressed source at different depths.

1) The above quantitative estimate of stress for domain reorientation as a function of depth; i.e. rocks are more responsive at depth;

2) The decrease in spontaneous magnetization with depth, being zero at the Curie-temperature depth;

3) The decrease in magnetic field intensity measured at the surface, as a source is placed deeper. For a focal zone modeled by a buried horizontal cylinder, the field falls off as $1/(\text{depth})^2$.

Figure 2 shows the relative magnitude of the net observed piezomagnetic anomaly to be expected. The curve is arbitrarily normalized to a reference depth of 5 km.

Again, the major field changes to be measured would originate from about the top 15 km of the lithosphere. It is this upper 15 or 20 km which is likely to be the most important in generating observable tectonomagnetic field anomalies for earthquake prediction.

REFERENCES

CARMICHAEL, R.S., Stress control of magnetization in magnetite and nickel and implications for rock magnetism, *J. Geomag. Geoelectr.*, **20**, 187–196, 1968.

CARMICHAEL, R.S., Depth calculation of piezomagnetic effect for earthquake prediction, *Earth Planet. Sci. Lett.*, **36**, 309–316, 1977.

GREEN, A.G., Interpretation of Project MAGNET aeromagnetic profiles across Africa, *Geophys. J.R. Astr. Soc.*, **44**, 203–228, 1976.

ISPIR, Y. and O. UYAR, An attempt in determining the seismomagnetic effect in N.W. Turkey, *J. Geomag. Geoelectr.*, **23**, 295–305, 1971.

Nagata, T., Application of tectonomagnetism to earthquake phenomena, *Tectonophysics*, **14**, 263–271, 1972.

Nulman, A., V.A. Shapiro, S. Maksimovskikh, N. Ivanov, J. Kim, and R.S. Carmichael, Magnetic susceptibility of magnetite under hydrostatic pressure and implications for tectonomagnetism, IAGA General Assembly, Seattle, Aug. 1977; *J. Geomag. Geoelectr.*, **30**, 585–592, 1978.

Rikitake, T., Geomagnetism and earthquake prediction, *Tectonophysics*, **6**, 59–68, 1968.

Yukutake, T. and H. Tachinaka, Geomagnetic variation associated with stress change within a semi-infinite elastic earth caused by a cylindrical force source, *Bull. Earthq. Res. Inst., Univ. Tokyo*, **45**, 785–798, 1967.

Magnetic Susceptibility of Magnetite under Hydrostatic Pressure, and Implications for Tectonomagnetism

A.A. NULMAN,* V.A. SHAPIRO,* S.I. MAKSIMOVSKIKH,*
N.A. IVANOV,* JOON KIM,** and R.S. CARMICHAEL***

**Institute of Geophysics, USSR Academy of Sciences, Sverdlovsk, U.S.S.R.*
***Gulf Energy and Minerals Company, Bakersfield, California, U.S.A.*
****Department of Geology, University of Iowa,*
Iowa City, Iowa, U.S.A.

(Received September 16, 1977)

The effects of hydrostatic pressure up to 2.7 kb on the susceptibility of magnetite single crystals, rocks and ore, and synthetic samples were studied. Susceptibility was measured using both static and alternating-frequency methods, with cyclic change of hydrostatic pressure and with time-dependent experiments. For a single crystal, susceptibility at saturation remanence decreases with pressurization, by about 25% for 2.5 kb, and is recovered reversibly upon release of pressure. For rocks and ore, susceptibility k_0 at small natural remanence increases slightly up to 0.2–0.8 kb, then decreases at higher pressures at rates up to -10%/kb. It recovers upon pressure release. For cyclic pressurization without a time lag between cycles, the initial increase is not observed. Annealing increases the value of k_0, and also suppresses the initial increase of k_0 with pressure. Time lags of hours between pressure cycles result in recovery of pre-pressurization values of susceptibility. This work is intended to assist in the interpretation of stress-induced changes observed for rocks in the earth's lithosphere. Applications are in understanding piezomagnetic field changes as precursors for earthquake prediction, and study of other secular tectonomagnetic changes.

1. Introduction

Physical properties, such as magnetization, of minerals and rocks depend on conditions of pressure and temperature in the lithosphere. Confining, or hydrostatic, pressure increases at a rate of about 320 bars/km. The magnetic properties of crystalline rocks are primarily due to magnetite. The magnetization is both induced, because of the minerals' susceptibility, and remanent. Directed stress, as from tectonic activity, changes the magnetization. This is due to inverse magnetostriction, and is the tectonomagnetic, or piezomagnetic, effect. Hydrostatic pressure can also produce tectonomagnetic effects and changes.

Piezomagnetic field variations are observed in seismically-active regions (MOORE, 1964; SMITH and JOHNSTON, 1976; SHAPIRO *et al.*, 1977a) due to changes in the focal-zone stress conditions, and in high-precision regional magnetic surveys repeated with time (SHAPIRO *et al.*, 1977b; SUMITOMO, 1977). Interpretation of these magnetic field variations would lead to understanding properties and behavior at depth, with application in such interests as seismotectonic conditions and earthquake prediction.

The effects of uniaxial stress on magnetic properties of minerals and rocks have been studied for some time (CARMICHAEL, 1968; OHNAKA and KINOSHITA, 1968; STACEY and JOHNSTON, 1972). Works on susceptibility include KAPITSA (1955), KERN (1961), NAGATA (1966), and KEAN *et al.* (1976).

However, little work has yet been done on the piezomagnetic effects of hydrostatic pressure (AVCHYAN, 1967; SHIVE, 1970; KIM, 1976), particularly for higher pressures. This paper is a progress report on work in this direction. The intent is to simulate the confining pressure condition in the upper lithosphere, and observe any changes in susceptibility and by inference the induced magnetization of crustal rocks. Magnetic susceptibility is a structure-sensitive property, reflecting the conditions of formation of the rock and subsequent mechanical deformation by tectonism. It can respond to applied stress with both reversible and irreversible changes, and an understanding of these will aid in relating tectonomagnetic variations to a geologic/stress model of the subsurface.

2. Experimental Procedure

The following experiments have been performed on magnetite-bearing samples:

2.1 Susceptibility at small natural remanent magnetization (NRM); i.e., k_0, at J_{NRM}

The samples (see Table 1, Nos. I–V) are spherical, with diameter of 7 mm. The magnetic grains are multidomain in size. The samples are covered with an adhesive film. The pressure chamber used is nonferromagnetic and nonconductive. The pressure medium is water, with the internal pressure determined by a hydraulic press manometer. Maximum pressure is 2.7 kb.

Table 1. Magnetite samples.

Sample group	Description	% Magnetite
I	Polycrystalline magnetite ore (0.094% TiO_2), grain sizes about 0.2×0.2 mm.	90
II	Magnetite ore (0.37% TiO_2), grain sizes from 0.2×0.2 to 0.5×0.5 mm.	76
III	Diorite with magnetite inclusions (0.63% TiO_2), grain sizes from .06×.06 to 0.3×0.3 mm.	24
IV	Diorite with magnetite inclusions (0.73% TiO_2), grain sizes from 0.1×0.2 to 0.3×0.3 mm.	19
V	Synthetic: magnetite (.02×.02 mm) in epoxy resin.	30
VI	Magnetite natural single crystal (1.06% TiO_2), cylinder of length 13.8 mm and diameter 5 mm; long axis parallel to [111] axis; coercive force=24 Oe.	100
VII	Synthetic: pure magnetite grains in epoxy resin (grain size .002–.15 mm, mode size=.008 mm), sample a cylinder of length 25 mm, dia. 8 mm.	1.7

The susceptibility is measured using a cylindrical inductor coil, enclosing the pressure chamber. The alternating-frequency (AF) method is used, with an alternating field of ± 0.7 Oe at 700 Hz, in an ambient field of 0.1 Oe. The samples measured have a small natural remanence. In the AF method, the frequency is measured for the coil empty and with the sample in it (CHECHERNIKOV, 1969). The susceptibility is

$$k=\frac{1}{4\pi C}\left(\frac{T^2}{T_0^2}-1\right)$$

where

T_0=period with no sample present

C=constant, depending on coil and sample geometries.

The calculated susceptibility is close to the true susceptibility k_0 for weak fields and low frequencies. For experimental conditions here, hysteresis and eddy current losses are absent. The period is measured with an accuracy of 10^{-7} using a frequency meter. A source of error in measuring the absolute value for k is the deviation of the sample shape from the spherical shape of a reference sample used to determine the constant 'C' above. The accurary of k_p/k_0 measurements is better; being affected only by the generator stability with time, and placement of the sample in the sensor space.

2.2 Susceptibity at saturation isothermal remanence; i.e., k_r, at $J_{r,sat}$

The samples are listed in Table 1, Nos. VI–VII. Number VII has grain sizes from single domain to multidomain. Samples are covered by a membrane. The pressure chamber is nonferromagnetic Be-Cu (modified after CARMICHAEL *et al.*, 1968). The pressure medium is kerosene and transformer oil, with the initial pressure determined from calibration using a phase transition of NH_4F. Maximum pressure used here is 2.5 kb.

The susceptibility is measured using cylindrical ballistic inductor coils, enclosing the pressure chamber. It is done by static detailing of the J vs. H relation using a field of ± 0.1 Oe, in a null ambient field. The samples have an initial saturation isothermal remanence.

3. Results

Figure 1 shows the relative changes of susceptibility, k_0, for the initial loading of samples I to V which have a small natural remanence. For pressures up to 0.2–0.6 kb, the susceptibility increases by a few percent. The k then decreases monotonically for higher pressures, up to 2.3 kb here. The rate of decrease ranges from very little, up to -9%/kb for the diorite samples. For Figs. 1 to 4, the hydrostatic pressure is determined by manometer.

In Fig. 1, the general trend of the curves is similar but there is variability between the samples. For example, the initial relative increase in k_0 ranges from $+1$ to $+8\%$. The variability is due to such factors as (and see Section 4):

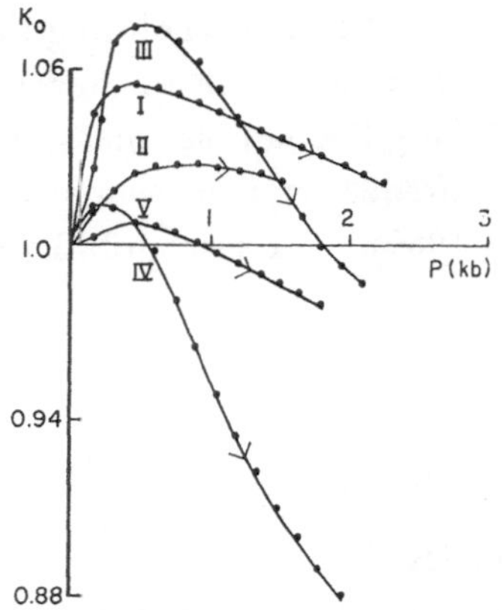

Fig. 1. Relative change of susceptibility for intial loading of rock samples (I to V) having a small natural remanence. Diameter of the points indicates measurement error in this experiment (Figs. 1–4).

i) Difference in magnitude and nature of initial natural remanence, to which the subsequent changes in k_0 are normalized,

ii) Variation in grain size, shape, and distribution; although all in Fig. 1 are multidomain,

iii) Variation in titanium content in the titanomagnetite,

iv) Condition of internal stress, for these natural rocks. This condition will be affected by the cyclic pressurization, to be done here.

Figure 2 gives results for cyclic compression for sample group I (magnetite ore). The first compression (curve with solid dots) is to 2.7 kb (point '1' along the abscissa). After an initial increase of k_0 up to 0.4 kb, k_0 decreases at a rate of about -3%/kb. When the pressure was removed, k_0 rebounds to point '1' on the ordinate; 8% higher than the original k_0. After a day's interval at room pressure, its value had declined to point '2' at 1.04 on the ordinate. After the second cycle of compression to 2.7 kb (point '2' along the abscissa), the k_0 is restored to point '2' at 1.08 on the ordinate. With another day's interval, the k_0 again declined to point '3' at 1.035. Cycles 3 and 4 extended to 1.2 kb and 0.4 kb respectively, and are traced out by point '3–3'

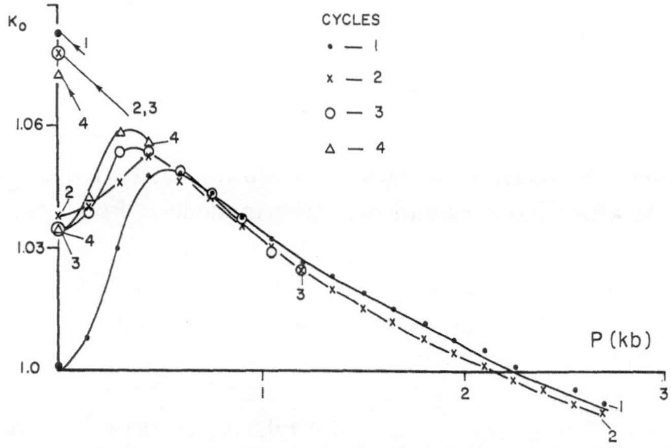

Fig. 2. Repeated compressions of sample group I. Cycles are denoted 1–4.

and '4–4.' Again, the pressurization resulted in an increase of a few percent in k_0 by 0.5 kb, then a decrease at about -3%/kb. Upon depressurization, the k_0 rebounded to a value that exceeded the cycle's initial value. For cycles beyond the first, for a given maximum pressure for a series of cycles, the unloaded k_0 was 5% above the final k_0 attained after relaxation.

It is seen that the cyclic pressurization of sample I has changed the general trend of the loading curve, reducing the initial increase of k_0. This makes the curve more similar to those of groups II, IV, and V for example (cf. Fig. 1).

The sample of group I was now annealed, for 3 hr at 250°C, in an inert atmosphere and null field. After this, k_0 had increased by 4%. Figure 3 shows the relative change of k_0 for various loading curves: loading to 2.7 kb before annealing (I), loading to 2.1 kb after annealing (II), and loading to 2.1 kb after a month's time interval (III).

Similar curves were obtained for samples from groups II–V. For example, Fig. 4 shows loading curves for sample IV (diorite) after annealing, with an intervening interval of 48 days. After the initial low-pressure increase, the k_0 decreases by -10%/kb.

Figure 5 shows the change in susceptibility at saturation isothermal remanence, k_r, for sample VI (magnetite natural single crystal). As the hydrostatic pressure increases to 2.2 kb, k_r decreases. The susceptibility, as measured along the [111] axis of the crystal, decreased by 23% for a pressure of 1.5 kb. The recovery is mostly reversible upon release of pressure.

Figure 6 shows a similar experiment for sample VII (synthetic powder dispersion). The decrease in k_r by 1.5 kb is about 33%. For these two samples (VI and VII), it thus appears that the powder sample, with grain sizes from single domain to multidomain, has yielded a greater effect than the multidomain single crystal. One might expect the opposite. An explanation is that the response of magnetization to stress depends in part on:

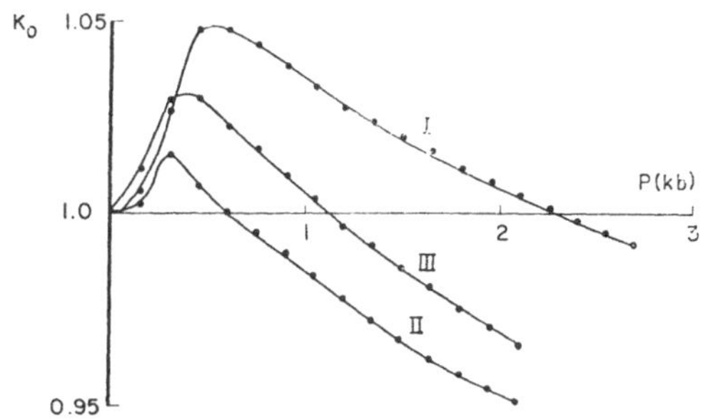

Fig. 3. Effect of annealing on sample of group I. Curve I before annealing, Curve II after, and Curve III after a month time interval.

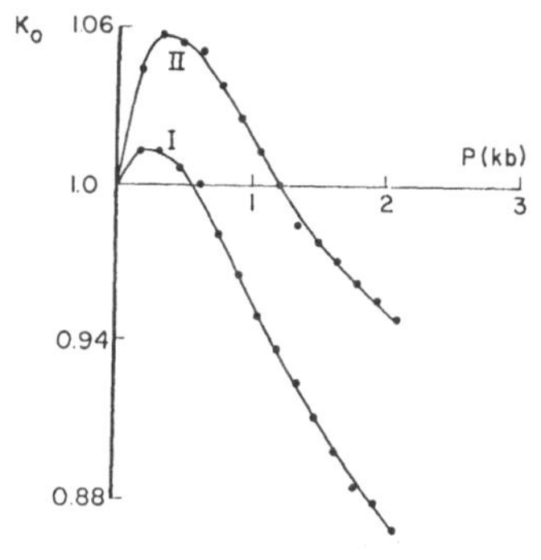

Fig. 4. Relative change of susceptibility for loading of sample IV after annealing (Curve I) and after an interval of 48 days (Curve II).

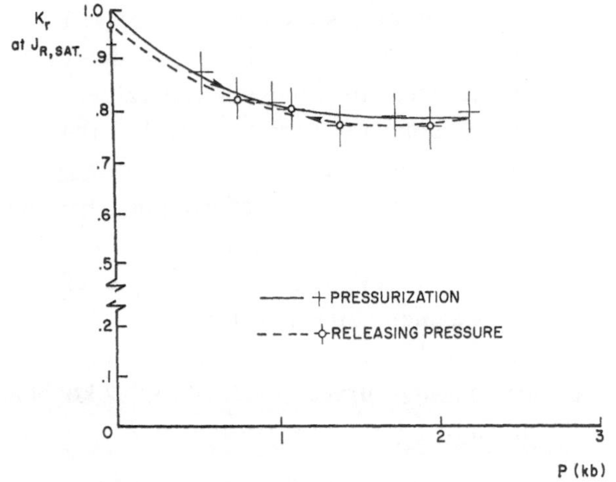

Fig. 5. Change in susceptibility at saturation remanence, for crystal sample VI.

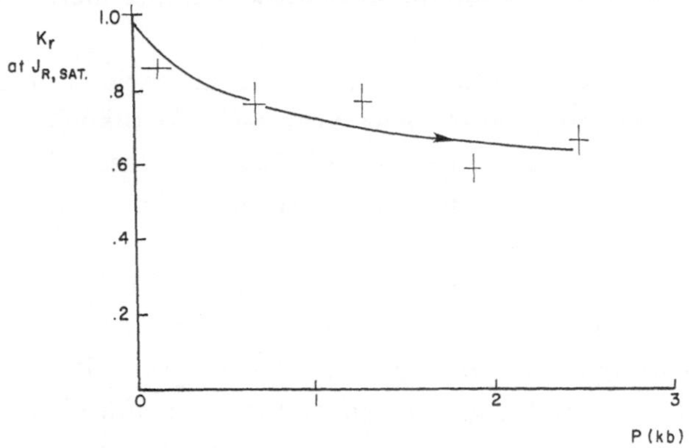

Fig. 6. Change in susceptibility at saturation remanence, for synthetic sample VII. Error bars indicate measurement error in this experiment.

i) Titanium content (e.g. KEAN *et al.*, 1976)—Neither sample has much titanium; the natural single crystal has 1.06% TiO_2, and the synthetic powder is pure magnetite.

ii) State of internal stress—Higher internal stress and local stress concentrations, would reduce the mobility of domains and walls, and thus reduce the responsiveness of magnetization to an influence such as externally applied stress. One measure of internal stress state, or magnetic 'hardness,' is coercive force. For the natural crystal, this is 24 Oe. For the synthetic powder sample, the bulk coercive force is less than this.

4. Discussion

Techniques have been developed to measure magnetic susceptibility of rocks to pressures of several kilobars. Such work should assist with interpretation of magnetic properties, and the resulting fields, of rocks in the crust and upper mantle.

The effect of pressure on physical properties can be important in considering the behavior of rocks 'in situ' in the lithosphere. It is also important in considering properties of rocks now sampled at the Earth's surface which have been subjected to pressure at depth at some earlier time in their geologic history. Susceptibility is a structure-sensitive property which depends on the mechanical condition of minerals and rocks, and changes in that condition such as due to burial or tectonism. It is a parameter which is influenced by both the external applied stress—hydrostatic or directed—and distribution of internal stresses.

As witnessed here for the magnetiferous rocks and polycrystalline multidomain magnetite carrying natural remanence, there are two ranges of effect under hydrostatic pressure. At low pressure (less than about 0.2–0.5 kb), susceptibility changes at a relatively high rate with pressure (although the absolute change is small); the magnitude of relative change is variable from sample to sample within a group or for different groups of rock; the change depends on the rate of loading or unloading and on the number of loading cycles; on time intervals between cycles; and on annealing. It is in this range for the rocks that the susceptibility is most sensitive to changes in the internal stress distribution, including the time-dependent effects of annealing (Fig. 3) and relaxation after unloading of a pressure cycle (Fig. 2).

The sensitivity of magnetization or susceptibility to stress depends in part on the state of internal stress, because the latter creates energy barriers that hamper domain rotation and especially domain wall displacement. For a set of rock samples of similar composition, texture, and remanence type, the magnetic behavior and responsiveness can be made reproducible and more similar by cyclic pressurization to some uniform hydrostatic pressure (cf. Figs. 1, 2). This is because the stressing generates a more comparable state of internal stress, other factors being equal, for the particular pressure level achieved. This makes more comparable the magnetic effects that depend in part on that internal stress condition.

At higher pressures (above 0.8 kb) for the rocks, and in the case of magnetite crystal samples with saturation remanence, the change in susceptibility with pressure is more uniform and similar. The susceptibility decreases, for the rocks at rates of up to $-10\%/\mathrm{kb}$. Upon unloading, the susceptibility changes more or less reversibly with respect to the 'high-pressure-range' behavior. The annealing did not change the nature of the susceptibility change (decrease) in this range.

Change in magnetization, such as susceptibility, caused by hydrostatic pressure can occur by domain wall motion and reorientation of the domain pattern in multidomain grains, and by rotation within single-domain grains or regions. Domain wall motion can occur when the magnetocrystalline potential barriers are removed or at least reduced by magnetostriction energy which is introduced by the applied hydrostatic pressure. The effects can be reversible and irreversible, depending on the magnitude of applied stress, and distribution of crystalline internal defects and stress concentrations.

Since susceptibility varies with hydrostatic pressure, the induced magnetization—as a consequence of the ambient earth's magnetic field—would vary as a function of depth of formation or burial. It is known that directed stress likewise produces changes in susceptibility. Understanding the nature and magnitude of these changes helps provide a basis for interpreting tectonomagnetic changes observed in seismically-active regions and in some precision repeated magnetic surveys. Because of the influence of hydrostatic pressure, there may also be a difference in susceptibility as measured for a rock under lab conditions of room pressure, and the true 'in situ' value for rocks at depth.

REFERENCES

AVCHYAN, G., Effect of hydrostatic pressure up to 8000 kg/cm² on various types of remanent magnetization of rocks, *Izv. Akad. Nauk USSR, Phys. Sol. Earth*, **7**, 70–76, 1967.

CARMICHAEL, R.S., Remanent and transitory magnetic effects of elastic deformation of magnetite crystals, *Philos. Mag.*, **17**, 911–927, 1968.

CARMICHAEL, R.S., A. SAWAOKA, and N. KAWAI, A multipurpose high-pressure microbomb, *Jpn. J. Appl. Phys.*, **7**, 1120–1124, 1968.

CHECHERNIKOW, V.I., *Magnetic Measurements*, pp. 139–143, Publ. House MGU, Moscow, 1969.

KAPITSA, S.P., Magnetic properties of erupted rocks under mechanical stresses, *Izv. Akad. Nauk USSR, Geophys. Ser.*, **6**, 489–504, 1955.

KEAN, W., R. DAY, M. FULLER, and V. SCHMIDT, The effect of uniaxial compression on the initial susceptibility of rocks, as a function of grain size and composition of their constituent titanomagnetites, *J. Geophys. Res.*, **81**, 861–872, 1976.

KERN, J.W., Effect of stress on the susceptibility and magnetization of a partially magnetized multidomain system, *J. Geophys. Res.*, **66**, 3807–3816, 1961.

KIM, J., Hydrostatic pressure effects on saturation remanent magnetization, susceptibility, and magnetic hardness of magnetite, Thesis, Michigan State University, 1976.

MOORE, G., Magnetic disturbances preceding the 1964 Alaska earthquake, *Nature*, **203**, 508–509, 1964.

NAGATA, T., Magnetic susceptibility of compressed rocks, I and II, *J. Geomag. Geoelectr.*, **18**, 73–80, 81–97, 1966.

OHNAKA, M. and H. KINOSHITA, Effect of uniaxial compression on remanent magnetization, *J. Geomag. Geoelectr.*, **20**, 93–99, 1968.

SHAPIRO, V.A., A.N. PUSHKOV, and K.N. ABDULLABEKOV, Geomagnetic investigations in the seismo-active regions of middle Asia, p. 90, IAGA Assembly, Seattle, Wash., Aug., 1977a.

SHAPIRO, V.A. and V.A. PJANKOV, Geomagnetic field secular variation anomalies and modern geodynamic processes in the Urals, p. 91, IAGA Assembly, Seattle, Wash., Aug., 1977b.

SHIVE, P., Deformation and remanence in magnetite, *Earth Planet. Sci. Lett.*, **7**, 451, 1970.

SMITH, B. and M. JOHNSTON, A tectonomagnetic effect observed before a magnitude 5.2 earthquake near Hollister, Calif., *J. Geophys. Res.*, **81**, 3556, 1976.

STACEY, F. and M. JOHNSTON, Theory of piezomagnetic effects in titanomagnetite-bearing rocks, *Pure Appl. Geophys.*, **V.97**, 146–155, 1972.

SUMITOMO, N., Secular variation anomalies and tectonomagnetism in Japan, p. 92, IAGA Assembly, Seattle, Wash., Aug. 1977.

Effect of Uniaxial Stress upon Remanent Magnetization: Stress Cycling and Domain State Dependence

Jacques Revol, Ron Day, and Michael Fuller

Department of Geological Sciences, University of California, Santa Barbara, California 93106, U.S.A.

(Received November 5, 1977)

Polycrystalline magnetite and rock samples have been subjected to uniaxial compression and stress cycling at room temperature. The changes in the components of remanent magnetization were recorded continuously as a function of stress, and the changes in direction and total intensity of magnetization were inferred. Different types of response were recognized, according to the type of magnetization the sample was carrying, i.e., high field or weak field remanence. The anomalous increase of weak field remanence previously reported, appears partly reversible under stress cycling. However, the changes in the lowest stress range are irreversible and reduce the zero stress magnetization from cycle to cycle. Reversible rotations of the magnetization vector of as much as 180° were observed during each half cycle and were primarily due to changes in the sign of the component of magnetization parallel to compression. An andesite whose magnetic phases are single domain according to hysteresis criteria showed a much smaller effect regardless of the type of magnetization it carried. These results again draw attention to the variety of stress responses and the importance of three component observations in field attempts to detect seismomagnetic precursors.

1. Introduction

The possibility of a seismomagnetic precursor arises because of the inverse piezo-magnetic effect exhibited by the magnetic minerals in stressed rocks. When changes in the magnetization of rocks take place at depth in the earth due to changes in stress, the magnetic ffeld at the earth's surface will be perturbed. Total field measurements of local geomagnetic field perturbations to an accuracy better than 0.1 γ (e.g., Smith and Johnston, 1976) are now possible, which is less than the field anomalies predicted by various models (Stacey and Johnston, 1972; Talwani and Kovach, 1972; Bhattacharyya, 1976). Magnetic anomalies associated with earthquakes in tectonically active areas have been reported by numerous authors (Breiner, 1964; Johnston *et al.*, 1975; Smith and Johnston, 1976), and although doubts remain concerning the demonstration of the seismomagnetic effect, some of these results appear well founded. An accurate estimate of the seismomagnetic effect in a particular area requires knowledge of the rocks at depth in the crust, of the stress build up prior to the earthquake and of the effect of stress upon the magnetization of magnetic phases present in the rocks. Our work is directed at obtaining a better understanding of this magnetic response of rocks to stress.

The effects of stress upon magnetic susceptibility and remanent magnetization have been widely investigated on rocks for stresses within the elastic range (KERN, 1961; NAGATA, 1970; POZZI, 1972; STACEY and JOHNSTON, 1972). It has been found that, in general, the susceptibility parallel to compression decreases while the transverse susceptibility tends to increase. However there is a variety of the response which is not explicitly considered by these studies, as shown by KEAN *et al.* (1976). Uniaxial stress tends to give a rotation of remanent magnetization away from the stress axis and a decrease in the total intensity of magnetization. However, the available data are inadequate to define such important aspects of the phenomenon as its again size or domain state dependence.

Until recently, very little work has been done involving stresses exceeding the elastic range. The major aim of our work is to document the magnetic behavior of rocks stressed throughout the entire range to failure. As a first step we have used small cylindrical samples, subjected to uniaxial compression at room temperature without confining pressure. For each sample, the stress was increased from zero until failure occurred. The effects of stress upon induced or remanent magnetization were monitored continuously. We recognize that this is an unrealistic experiment in terms of simulation of the failure likely in the crust, but in terms of interpretation of magnetic effects it is an important preliminary to the main investigation.

The most important conclusion to be drawn from our earlier study (REVOL *et al.*, 1977) is that magnetic effects of uniaxial compression up to failure are profound and readily observed in polycrystalline and in some rock types. Throughout the range of stress, substantial changes in remanence and susceptibility were seen. In the lower stress range there is a change in stress sensitivity, which may be related to the onset of dilatancy. There were also more erratic effects in the approach to failure. The appearance of the latter allowed us to predict failure systematically in the laboratory for those samples that showed any magnetic response to stress.

The application of stress caused changes in the components of both remanent magnetization and induced moment, with changes in the component parallel to compression dominating the response. Changes in the intensity and direction of the magnetization vector are related to the initial orientation of magnetization, as well as to the type of magnetization the sample is carrying. In the case of strong field remanent magnetization (IRM_s) the total intensity of magnetization decreased monotonically with increasing stress in zero field. In contrast, demagnetized samples, and samples carrying a weak field thermo-remanent magnetization (TRM), showed an anomalous increase in total intensity with increasing stress in zero field, due principally to the increase parallel to compression. This effect produced large changes in the direction of the magnetic moment (Fig. 1). The magnitude of this increase, and the extent of the stress range over which it occurs appear to be related to be the magnitude and direction of magnetization before stressing. A serpentinite sample carrying natural remanent magnetization also showed this anomalous increase in intensity under stress in zero field.

The stress effect upon susceptibility was earlier shown to be strongly dependent

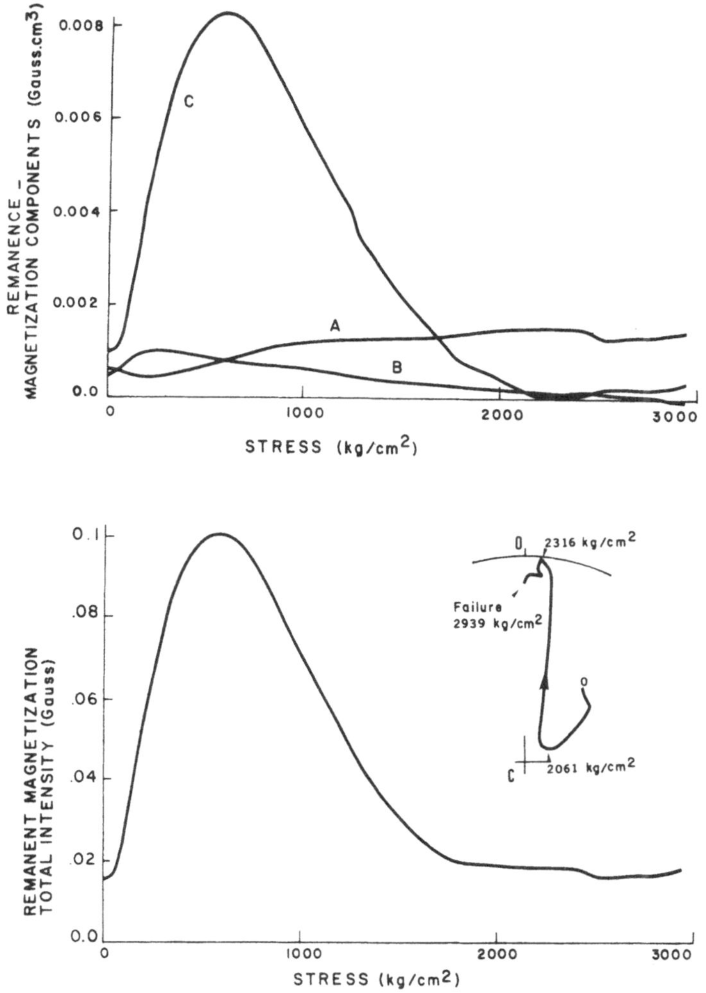

Fig. 1. Components, intensity and rotation of remanent magnetization under stress for multidomain magnetite carrying a TRM.

upon grain size (KEAN *et al.*,1976). Thus synthetic samples of coarse grain magnetite (150–175 μm) gave stress sensitivities of $8\times10^{-4}\,cm^2\cdot kg^{-1}$, while 1–2 μm dispersions gave 0.3×10^{-4}. Synthetic titanomagnetites showed an even greater variation from 4×10^{-3} to $0.3\times10^{-4}\,cm^2\cdot kg^{-1}$. Rock samples showed less variability with values ranging from 6×10^{-4} to $1\times10^{-4}\,cm^2\cdot kg^{-1}$, with the coarse grain titanomagnetite-bearing samples giving the larger value. These values are comparable with those cited by STACEY and JOHNSTON (1972).

The stress sensitivity for remanence was also given by STACEY and JOHNSTON (1972) and values of between 0.7×10^{-4} and $18.7\times10^{-4}\,cm^2\cdot kg^{-1}$ were cited with rocks with the more titaniferous magnetite being more sensitive. The discovery of the anomalous increase in remanence parallel to compression, in polycrystalline magnetite and rocks, carrying weak field remanence, complicates the description of the stress sensitivity of remanence considerably. If calculated from $M_{(\sigma)}/M_0 = 1/(1+S_M\sigma)$, S_M changes sign and magnitude according to the stress range, with a maximum value of 3×10^{-3} cm^2·kg^{-1} between 0 and 300 bars. It appears that the slope at a particular stress, normalized by the magnetization at that stress, $(1/M_{(\sigma)})(\partial M/\partial\sigma)$, is a more meaningful form and is equivalent to that normally quoted for the low stress range. The values for this stress sensitivity parameter vary from $+3\times10^{-3}$ cm^2·kg^{-1} through zero to a comparable negative value (Fig. 1). Although not as large in rock samples, similar effects have been seen, e.g., in serpentinite.

The presence of the anomalous increase parallel to compression means that the stress sensitivity of remanence can be considerably larger than that of susceptibility. For example, in the polycrystalline magnetite the maximum susceptibility sensitivity of 4×10^{-4} is almost an order less than the maximum remanence sensitivity. Before we can translate this into an assessment of the relative importance of the seismomagnetic effect, the roles of susceptibility and remanence in determining a magnetic field anomaly must be defined by the Koenigsberger ratio

$$Q = \frac{\mathrm{NRM}}{k\,H_{\mathrm{earth}}} \approx \frac{\mathrm{NRM}}{k\,0.5}\,.$$

Thus, for the polycrystalline magnetite with a typical value of TRM of 0.05 Gauss and a susceptibility of 0.15 Gauss/Oe, the remanent effect will dominate over the stress ranges for which the anomalous increase grows and decays. In the case of the serpentinite the remanence effect will also dominate.

In this paper, we describe additional experiments primarily related to cycling, to determine if the anomalous increase in total intensity of remanent magnetization (usually due to an increase in the component parallel to compression) is reversible. We also distinguish between fine-particle and multidomain behavior under stress.

As in our earlier experiments we have not made simultaneous measurements of strain and components of magnetization, nor have we introduced a confining pressure.

2. Experimental Procedure

Our experimental technique allows for the continuous measurement of the changes in the three components of magnetization vs. applied uniaxial stress. The equipment and procedures have been described earlier (REVOL *et al.*, 1977). Modifications have been introduced to speed up the process of data reduction.

The output signals from the magnetometer and pressure transducer are now sent to an analog to digital converter, the output of which is connected to a teletype. A reading of the four parameters can be made approximately every half second, but averages can be taken over a variable number of readings to give quasi-continuous recording.

The data are then reduced to obtain the total intensity of magnetization vs. stress, and the directional changes of the vector are plotted on a stereonet. The axis of compression (C axis) is perpendicular to the plane of the projection of the net. The initial azimuthal direction is arbitrary and only the relative changes in direction are of interest.

3. Magnetic Characteristics of Samples

The rock samples used in these experiments were in the form of small cylinders with a length to diameter ratio of about 1.29, a value around which the most uniform stress distribution is obtained. This also gives a shape anisotropy that should not exceed 10%.

The following rock types were used:

1) Natural polycrystalline magnetite (Norway).
Diameter=0.43 cm, Length=0.57 cm.
Saturation magnetization $J_s=95.0$ emu/g,
Remanent magnetization $J_r=1.1$ emu/g,
$J_r/J_s=0.01$, coercive force $H_c=17$ Oe,
Remanent coercive force $H_{rc}=145$ Oe, $H_{rc}/H_c=8.5$.
The ratios H_{rc}/H_c and J_r/J_s indicate multidomain behavior for this material, also suggested by the shape of the A.F. demagnetization curve.

2) Porphyritic andesite from the margin of a dike.
Wildwood Park, California.
Diameter=0.43 cm, Length=0.56 cm.
$J_s=2.77$ emu/g of sample, $J_r=1.72$ emu/g of sample,
$J_r/J_s=0.62$, $H_c=323$ Oe, $H_{rc}=400$ Oe, $H_{rc}/H_c=1.24$.
These values indicate that this rock is mainly composed of single domain magnetic carriers. The Curie point is 340±10°C indicating a 0.4x value titanomagnetite. The opaques are about 2 μm or less in diameter.

4. Results

4.1 Stress cycling

In this section we describe the series of experiments designed to determine if the previously reported anomalous increase in magnetization (Revol *et al.*, 1977) is reversible for different remanent magnetizations.

4.1.1 Magnetite carrying isothermal remanent magnetization (IRM)

The application of uniaxial stress in zero field produced a decrease in each of the three components of magnetization. The total intensity of magnetization was reduced by an order of magnitude during the first cycle. The rate of decrease was high at low stresses but decreased progressively in the range corresponding to half the breaking stress. This type of behavior has already been observed in the elastic range (Nagata, 1969), but evidently it also applies when the sample is taken near to failure. The behavior of the three components, intensity and rotation of the

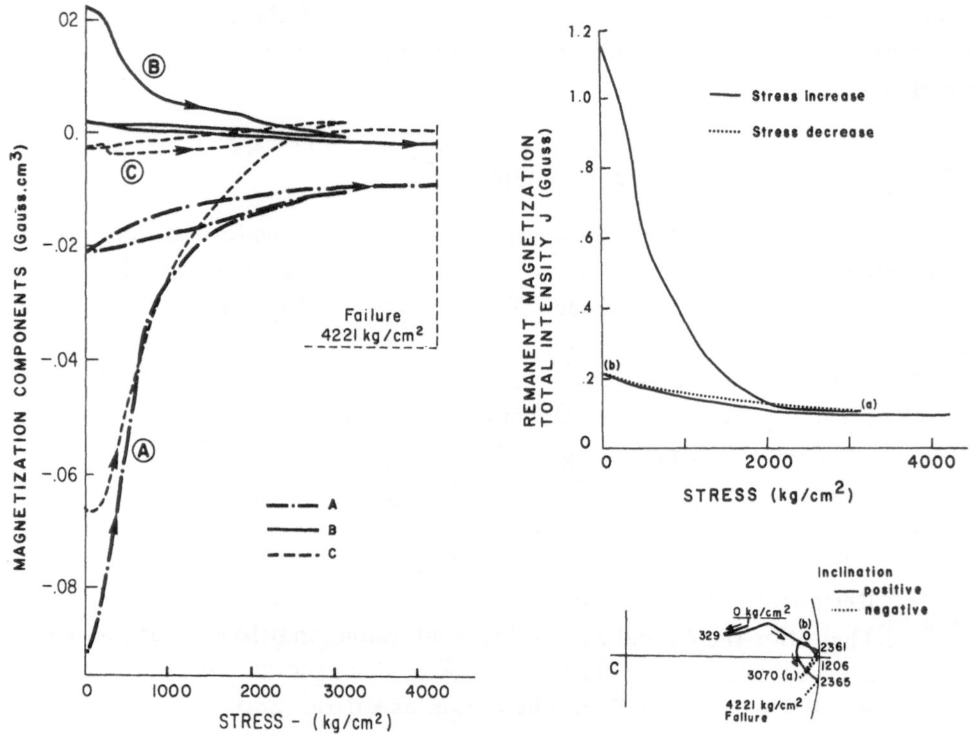

Fig. 2. Changes in the components of remanent magnetization due to stress cycling between 0 and 1,200 kg/cm² for multidomain magnetite carrying an IRM.

magnetization vector is shown in Fig. 2. Stressing produced changes of about 45° in both declination and inclination.

It is worth mentioning that in this case the sample was cycled to a high stress value, resulting in an increase of the ultimate failure stress. This can be expected, and is probably due to cold working of the sample.

4.1.2 Magnetite carrying thermo-remanent magnetization (TRM)—Intermediate stress range cycling

A sample was given a TRM by heating to 600°C in a vacuum and cooling in a 0.4-Oe field.

This sample was cycled 5 times from 0 to approximately 1,200 kg/cm² and then stressed to failure. The changes in the three components of magnetization brought about by the applied stress are shown in Fig. 3a. The plot of total intensity vs. stress and the directional changes of the magnetization vector are given in Fig. 3b. The numbers on the curves in Figs. 3a and 3b indicate the successive stress cycles. The irreversible effect of stress on the *A* component, which was perpendicular to compression, was to reduce its value at zero stress from cycle to cycle. During a particular stress cycle, the *A* component decreased in value with the application of stress, and passed through zero without any apparent change in behavior.

Cycling produced an essentially reversible increase in the *B* component, which was also perpendicular to compression, but a decrease was observed before the sample reached failure.

Under stress, the *C* component, which was parallel to compression, decreased in absolute value, passed through zero magnetization and continued to increase in the opposite sense with further pressure. After each cycle, the magnetization at zero stress was progressively reduced in absolute value. It eventually changed sign after the fourth cycle. During the fifth cycle the pressure was released too fast to permit accurate determination of the magnetization between 1,200 and 0 kg/cm². The large offset observed between points 9 and 10 of Fig. 3a could therefore be an artifact but could also be explained by cracking.

Figure 3b reveals a decrease in total intensity of about 75% of the initial value during the first cycle, followed by an increase to approximately the original value. The minimum was observed at a stress corresponding to one-third of the breaking stress. Further cycling reduced progressively the zero stress magnetization by about 90% of the initial value, but the subsequent increase was predominantly reversible. The questionable offset between points 9 and 10 prevents any definite conclusion about the last stress cycle; however it appears that the magnetization reached a maximum greater than the initial intensity, when the sample was stressed to failure.

Directional changes were largely reversible for at least the first three cycles, with little azimuthal deviation. The magnetization vector rotated away from the compression axis, and passed through the plane perpendicular to compression, which is the plane of projection of the stereonet (Fig. 3b). It then moved back towards the axis of compression, to a direction opposed to the initial vector for a stress of about 1,200 kg/cm². When the stress was released the path was retraced. During stress release on the fourth cycle the vector rotated back by only about 80°, and followed the same trajectory in reverse when the stress was increased again.

4.1.3 *Magnetite carrying TRM—Low stress range cycle*

This sample was given TRM under the same conditions as described before. It was cycled between 0 and 400 kg/cm², and the changes in total intensity and the directional changes are given in Fig. 4.

The first three cycles reduced the zero stress magnetization. The corresponding directional changes were reversible. The stress was then increased to about 900 kg/cm², which resulted in rotation of the magnetization vector away from the compression axis and switched the sign of the inclination. The release of stress gave a zero stress magnetization less than half the initial value (points 3 and 4). Three more cycles were carried out between 0 and 400 kg/cm² (points 4 and 5). The changes in intensity as well as the directional changes were reversible during these cycles, with the magnetization vector alternating between a positive and a negative inclination.

The last stress increase produced a large increase in magnetization to give a value almost identical to the initial value at a stress of about one-third of the breaking stress. A comparable decrease in intensity was observed before the sample reached failure. The corresponding directional change involved rotation of the

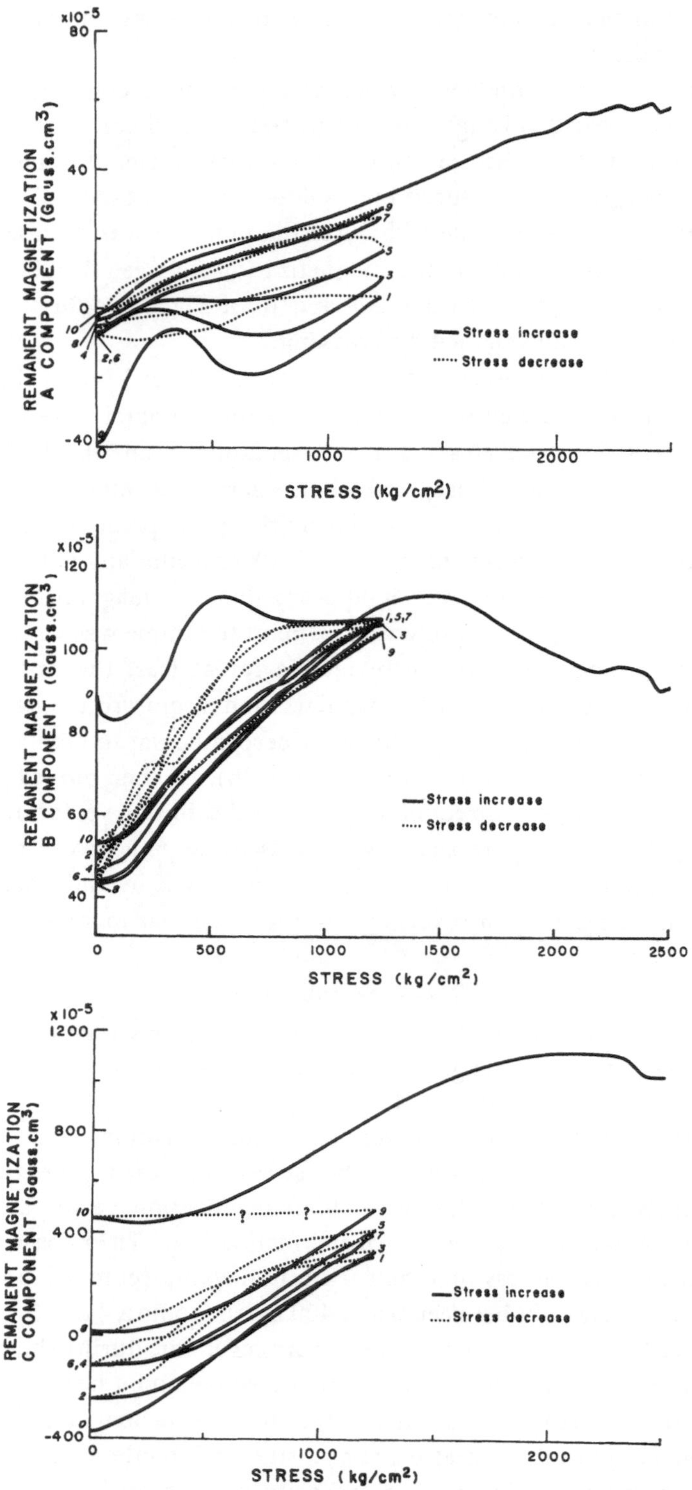
x10^-5
80
40
0
-40
REMANENT MAGNETIZATION
A COMPONENT (Gauss.cm³)
0
1000
2000
STRESS (kg/cm²)
Stress increase
Stress decrease
x10^-5
120
100
80
60
40
REMANENT MAGNETIZATION
B COMPONENT (Gauss.cm³)
0
500
1000
1500
2000
2500
STRESS (kg/cm²)
Stress increase
Stress decrease
x10^-5
1200
800
400
0
-400
REMANENT MAGNETIZATION
C COMPONENT (Gauss.cm³)
0
1000
2000
STRESS (kg/cm²)
Stress increase
Stress decrease

Fig. 3a

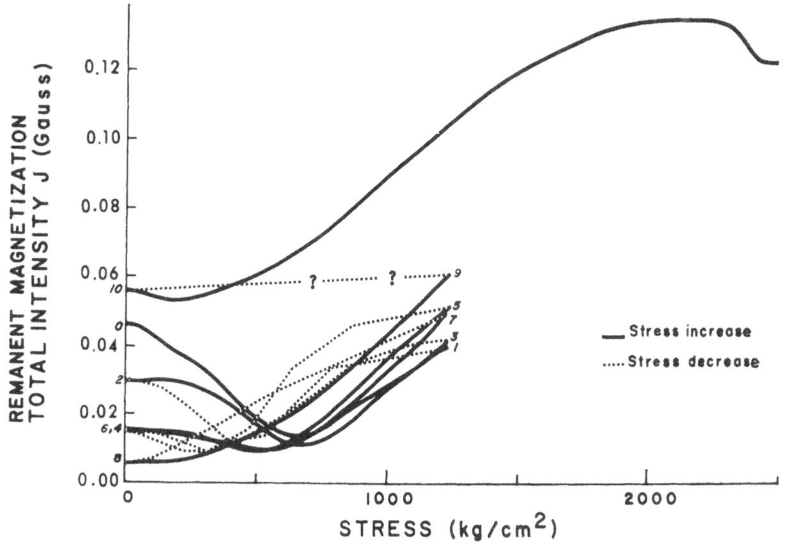

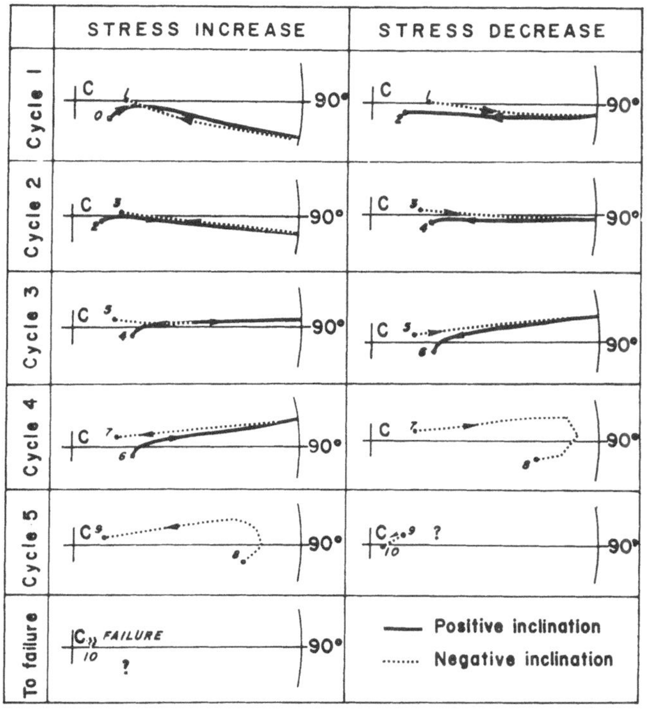

Fig. 3b

Fig. 3a. Changes in the components of remanent magnetization due to stress cycling between 0 and 1,200 kg/cm² for magnetite carrying a TRM.

Fig. 3b. Changes in total intensity of remanent magnetization and directional changes due to stress cycling between 0 and 1,200 kg/cm² for magnetite carrying a TRM.

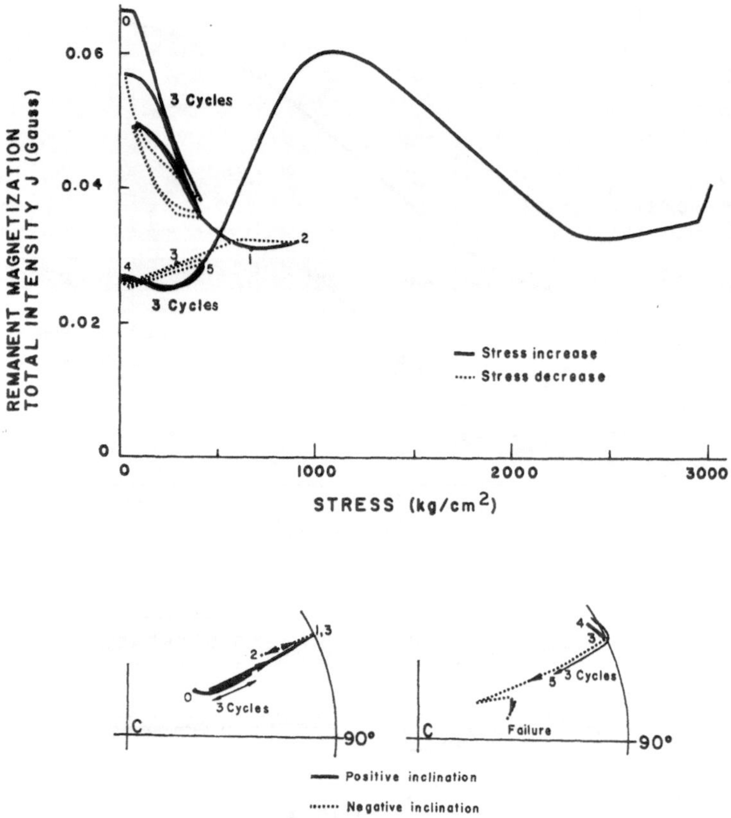

Fig. 4. Changes in total intensity of remanent magnetization and directional changes due to stress cycling between 0 and 400 kg/cm² for magnetite carrying a TRM.

magnetization vector toward the axis of compression in an opposed sense to that of the initial orientation.

4.1.4 Magnetite carrying TRM—Cycling between 1,000 and 1,200 kg/cm²

This sample carried a TRM similar to those previously described, and was cycled in an intermediate stress range. For stresses lower than the cycling range, all three components of magnetization initially decreased in absolute value. For stresses exceeding the cycling range, the component parallel to compression and one of the components perpendicular to it, increased in absolute value, whereas the second component perpendicular to compression kept decreasing in absolute value. None of the components actually crossed the zero magnetization value, contrary to cases described above.

The intensity of magnetization decreased by 60% of the initial value and started to increase for stresses just below the stress cycling range. Cycling five times gave approximately reversible changes in magnetization, producing only a slight decrease from cycle to cycle. For stresses exceeding the cycling range in the final approach to failure, the intensity increased again before the appearance of an erratic behavior announcing failure. The magnetization rotated first away from the compression axis, and then back toward it at stresses higher than the cycling interval.

This sample did not show, in the lower stress range, the large increase observed in other samples carrying the same type of magnetization (TRM). However a smaller increase in magnetization was observed at higher stresses. It may actually be the same effect, shifted with respect to stress by inhomogeneity of internal stress.

4.2 Grain size control of response to uniaxial stress

In our earlier paper (Revol *et al.*, 1977) we reported that a granodiorite sample containing fine magnetic carriers with saturation IRM showed virtually no change

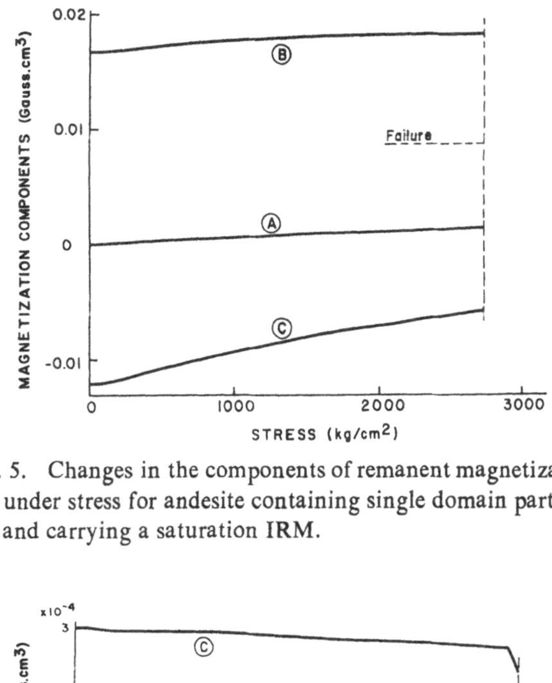

Fig. 5. Changes in the components of remanent magnetization under stress for andesite containing single domain particles and carrying a saturation IRM.

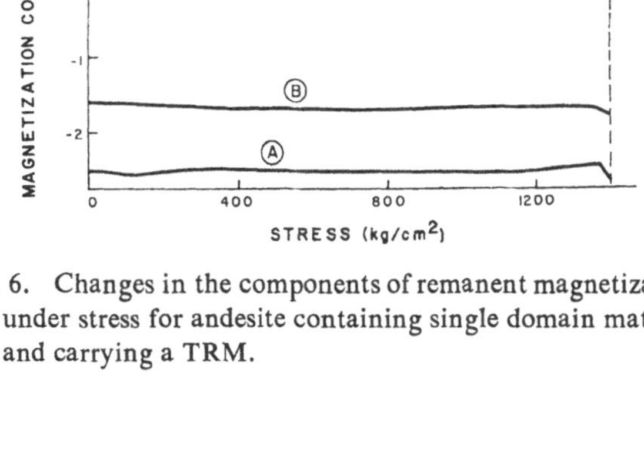

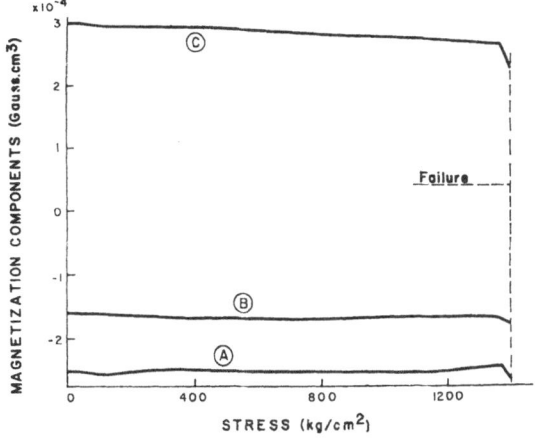

Fig. 6. Changes in the components of remanent magnetization under stress for andesite containing single domain material and carrying a TRM.

under stress. Stress cycling was also studied on this sample and showed no significant effect. Additional experiments discussed below have been conducted with samples containing fine magnetic carriers.

The andesite described above and containing fine titanomagnetite particles carrying a saturation IRM showed only a small effect due to the applied stress (Fig. 5). The components of magnetization perpendicular to compression increased slightly in absolute value while the component parallel to compression decreased. The total intensity of magnetization decreased very little and the magnetization vector slightly rotated away from the compression axis.

The andesite was given a TRM under the same conditions as magnetite. The plot of the components of magnetization vs. stress is given in Fig. 6. The component parallel to compression decreased under stress but no noticeable change was observed in the components perpendicular to compression. The total intensity of magnetization showed a small decrease at the approach to failure, and no change in direction was observed.

5. Conclusions

The present paper establishes the degree of reversibility of the previously reported anomalous increase in TRM of polycrystalline magnetite under stress cycling, in zero field and at room temperature. It is clear that whereas the changes in the lowest stress range are irreversible and tend to reduce the magnitude of the zero stress magnetization after cycling (also reported by HENYEY *et al.*, 1978), the anomalous increase which occurs in a somewhat higher stress range is reversible. The importance of this result is that the effect may be expected to persist throughout many natural cycles. In multidomain magnetite this anomalous response of remanent magnetization will be greater than the changes in susceptibility.

The distinction between the response of multidomain and fine grain single domain magnetic material is remarkable. The latter shows very small responses in both susceptibility and remanent magnetization. For example, as we noted in the introduction, the earlier work of KEAN *et al.* (1976) suggests values of order 10^{-5} $cm^2 \cdot kg^{-1}$ for the stress sensitivity for susceptibility. The results from this paper suggest that if a similar sensitivity parameter is used, the values for TRM carried by single domain material will also be of order 10^{-5}.* In contrast, the multidomain remanence effect is initially two orders larger in the polycrystalline magnetite and appears long before failure. Moreover, the form of the effect does not actually permit useful definition of the stress sensitivity in the manner used previously. Both in the polycrystalline magnetite and in rock samples containing multidomain magnetite the remanent moment effect dominates over the susceptibility effect in the earth's field. In fact, in comparing the stress effect in the presence and absence of a field we have found little difference. Thus, in the polycrystalline magnetite we have

* This very small stress effect upon weak field remanence of single domain material suggests that only in extreme circumstances will the stable NRM of rocks be affected by stress.

studied, the magnetic response to a stress of kilobars unconfined uniaxial compression is dominated by the soft multidomain material. If such material carries a weak field NRM or TRM it is the anomalous increase parallel to compression which is the principal effect. It now remains to find out how general this phenomenon is, particularly if it is seen when the sample is confined, and if so, whether it can eventually be used as a seismomagnetic precursor.

This work was supported by a grant from the U.S. Geological Survey. The authors are grateful to V. Sefcik for interfacing the equipment with a mini-computer.

REFERENCES

BHATTACHARYYA, B.K., Development of an Optimized Experiment by Measuring Piezomagnetic Signals of Geological Origin, Space Sciences Laboratory Series 16, Issue 27, June 2, 1976.

BREINER, S., Piezomagnetic effect at the time of local earthquakes, *Nature*, **202**, 790–791, 1964.

HENYEY, T.L., S.J. PIKE, and D.F. PALMER, On the measurement of stress sensitivity of NRM using a cryogenic magnetometer, *J. Geomag. Geoelectr.*, **30**, 607–617, 1978.

JOHNSTON, M.J.S., G.D. MYREN, N.W. O'HARE, and J.H. RODGERS, A possible seismomagnetic observation on the Garlock fault, California, *Seismol. Soc. Am. Bull.*, **65**, 1129–1132, 1975.

KEAN, W.F., R. DAY, M. FULLER, and V.A. SCHMIDT, The effect of uniaxial compression on the initial susceptibility of rocks as a function of grain size and composition of their constituent titanomagnetites, *J. Geophys. Res.*, **81**, 861–872, 1976.

KERN, J.W., The effect of stress on the susceptibility and magnetization of a partially magnetized multidomain system, *J. Geophys. Res.*, **66**, 3807–3816, 1961.

NAGATA, T., Basic magnetic properties of rocks under the effects of mechanical stresses, *Tectonophysics*, **9**, 167–195, 1970.

POZZI, J.P., Piezomagnetic properties of hematite and effect of reheating, *Ann. de Geophysique*, **28**, 1–13, 1972 (in French).

REVOL, J., R. DAY, and M. FULLER, Magnetic behavior of magnetite and rocks stressed to failure—Relation to earthquake prediction, *Earth Planet. Sci. Lett.*, **37**, 296–306, 1977.

SMITH, B.E. and M.J.S. JOHNSTON, A tectonomagnetic effect observed before a 5.2 earthquake near Hollister, California, *J. Geophys. Res.*, **81**, 3556–3560, 1976.

STACEY, F.D. and M.J.S. JOHNSTON, Theory of the piezomagnetic effect of titanomagnetite-bearing rocks, *Pure Appl. Geophys.*, **97**, 146–155, 1972.

TALWANI, P. and R.L. KOVACH, Geomagnetic observations and fault creep in California, *Tectonophysics*, **14**, 245–256, 1972.

On the Measurement of Stress Sensitivity of NRM Using a Cryogenic Magnetometer

T.L. HENYEY,* S.J. PIKE,* and D.F. PALMER**

*Department of Geological Sciences, University of Southern California, Los Angeles, California, U.S.A.
**Department of Geology, Kent State University, Kent, Ohio U.S.A.

(Received December 13, 1977)

Preliminary stress sensitivities of NRM of two rocks from along the San Andreas fault, California, have been investigated using a cryogenic magnetometer and uniaxial pressure vessel. Of particular interest has been the irreversible behavior in response to cycling in the stress range 0–500 bars. The effect of hydrostatic pressure has also been examined. Sensitivities parallel to the compression axis behave as predicted by simple theory and show irreversible behavior during initial stress cycles. In contrast, transverse sensitivities show significant deviations from simple theory. Stress sensitivities of NRM do not appear to be appreciably affected by increase in hydrostatic pressure up to 0.5 kb.

1. Introduction

The effect of tectonic stress on rock magnetism has been of interest for many years (see for example NAGATA, 1969; also more recently JOHNSTON and STACEY, 1969; YAMAZAKI and RIKITAKE, 1970; SMITH and JOHNSTON, 1976; ISPIR and UYAR, 1971). The suggestion that precursory changes in stress may accompany earthquakes has broadened interest in tectonomagnetism, with the possibility that field magnetometers can be used as stress transducers to predict earthquakes.

Laboratory measurements of piezomagnetism in materials of geologic interest can probably be credited to Russian workers (e.g., KALASHNIKOV and KAPITSA, 1952; GRABOVSKII and PARKHOMENKHO, 1953). Subsequent work by others (KERN, 1961; STACEY, 1962, 1964; NAGATA and KINOSHITA, 1965; OHNAKA and KINOSHITA, 1968; NAGATA, 1966a, b, 1970a, b; STACEY and JOHNSTON, 1972) verified and expanded upon the earlier studies, and provided estimates of expected tectonomagnetic effects based upon the laboratory measurements of stress sensitivities.

In pioneering work on piezomagnetism, samples were sequentially stressed and unstressed in an applied field and zero field, and after each stress cycle, the magnetization was measured at zero pressure. Measurements could then be made after successively higher peak stresses had been attained to provide estimates of the interdependence of stress and magnetization on a quasi-continuous basis.

With the development of highly sensitive cryogenic magnetometers, it has become possible to significantly increase the sophistication of piezomagnetic measurements. Magnetization along three orthogonal axes as a function of stress can be measured continuously in an applied or zero field. Specimen strains (longitudinal and circumferential) can also be measured continuously during a piezomagnetic experiment. Superconducting shields are highly effective in attenuating ambient magnetic field fluctuations, additionally improving signal to noise ratios, as well as providing highly stable 'trapped' fields.

As noted by Kean *et al.* (1976), the posssible magnetic changes which might be expected upon the application of stress to a rock include:

1) Change in susceptibility (χ),
2) Change in the natural remanent magnetization (NRM),
3) Acquisition of a piezomagnetic remanent magnetization (PRM).

Nagata (1969) points out that for low stresses (less than 500 bars), 1) and 2) above exhibit largely reversible behavior, while 3) is irreversible. However, for soft NRM components or high stresses, irreversible behavior in 2) can occur; this is the well-known 'stress-demagnetization' effect (Nagata, 1970a). The cryogenic magnetometer is capable of investigating each of the three effects above, independently, or in combinations, and along three orthogonal directions. For example, both NRM and χ changes are observed in the presence of a trapped field, while only NRM changes will occur in a zero field environment. A demagnetized sample stressed in the presence of an ambient field will provide primarily sensitivity of susceptibility. (We thank the reviewers for noting that sensitivity of susceptibility can also be gotten from two successive experiments in a constant ambient field in which the sample has been flipped by 180°.) The application of hydrostatic pressure is also a straightforward process in the cryogenic magnetometer.

In order to isolate each of the different piezomagnetic effects, and to determine their relative magnitudes, measurements should be made on the same specimen. This is only practical in the low stress range, where strains are reversible. It is generally not possible where dilatancy, creep, or fracture occurs. In these cases, using closely spaced specimens from a single sample is the best compromise.

In this paper, we report on some measurements of the reversibility of NRM of two igneous rocks in the low stress range (0–500 bars). Our samples come from rock bodies near the San Andreas fault system in central and southern California (Fig. 1).

Fig. 1. Sample localities.

Except possibly for rocks adjacent to the rupture surface or near material 'singularities,' stress changes during strain buildup and seismic slip may not exceed a few hundred bars (BRUNE and ALLEN, 1967; BRUNE *et al.*, 1969).

2. Sample Characterization

Two distinctly different samples are discussed here. Sample 760S001 is a basalt from central California obtained by B. Smith and M. Johnston of U.S. Geological Survey and taken from an outcrop using a hand-held core drill. Sample AN-2-725 is a quartz monzonite from a deep drill core (725 feet) in the southern California batholith. Both are representative of large rock bodies found near the San Andreas fault system.

Detailed petrographic and mineralogic analysis of sample 760S001 has not yet been made. AN-2-725 is medium-grained with no gneissose or schistose structure, and no evidence of surface related weathering processes. The core from which it was taken showed high mineralogical uniformity. Some minor granulation and microfracturing is found throughout the specimen. Microfractures are healed by solution-overgrowths leading to the development of sutured contacts and lobate borders in quartz grains, and by minute, generally monomineralic veinlets of quartz, magnetite, hematite, or pyrite along the microfractures. Total oxide content is about 0.1%. Examination of AN-2-725 shows three distinct generations of titanomagnetite with varying grain sizes and grain shapes. The primary titanomagnetite has grain sizes ranging from 0.01 to 0.3 mm. The large grains are subhedral to anhedral and are often associated with ferromagnesian silicates. The small grains are nearly perfect octahedra enclosed poikilitically in feldspars. These grains were apparently included in the feldspars early in the crystallization of the rock and were thus isolated, inhibiting further growth. While these grains are too small to be reflected in a standard modal analysis, they may constitute as much as 10% of the total mass of early magnetite in the rock. A second generation of magnetite has formed through granular exsolution of ilmenite from the earlier titanomagnetite. This secondary magnetite is homogeneous and the ratios of exsolved ilmenite and magnetite suggest that the maximum amount of ulvospinel in the titanomagnetite did not exceed 10%. A third generation of magnetite occurs as thin veinlets in microfractures. Preliminary X-ray diffraction measurements suggest at least two compositions of magnetite within the rock.

Specimens were machined into cylinders 2.3 cm long $\times$ 2.3 cm diameter with parallelism and flatness to ± 0.003 cm. Edges were chamfered to minimize the chance of flaking.

3. Experimental Apparatus

Good reviews of superconducting magnetometers and their application to problems of geologic interest have been presented elsewhere (CLARKE, 1973, 1976;

LOUNASMAA, 1974; GOREE and FULLER, 1976), and will not be elaborated upon here. Our instrument contains a 6.4 cm 'hot hole' access tube with two sensing coils (one axial and one transverse), which are flux-coupled to two 'DC' SQUIDS. The specimen can be rotated 90° to obtain the third orthogonal component. For simplicity, during one experimental run on a given sample, we normally orient the transverse component of the NRM vector parallel to the transverse pickup coils, and in a succeeding run on the same sample, the transverse component is oriented perpendicular to the coils. This procedure poses limitations only in the case of strongly irreversible behavior.

A versatile pressure vessel (Fig. 2) has been constructed which permits the application of hydrostatic pressure, simultaneous measurement of strain, and the study of relatively large specimens (cylinders up to 3.8 cm long $\times$ 2.5 cm in diameter). An air to oil intensifier supplies continuous change in axial pressure up to 6 kb on a 1.5 cm^2 cross section, and up to 1 kb hydrostatic pressure through two separately metered lines to the pressure vessel. The pressure vessel is constructed from age-hardened BeCu with a net magnetic moment less than 10^{-3} emu and no discernible piezomagnetic response at levels relevant to our measurements. Volumetric strain is determined using axial and circumferential strain gages affixed to the specimen.

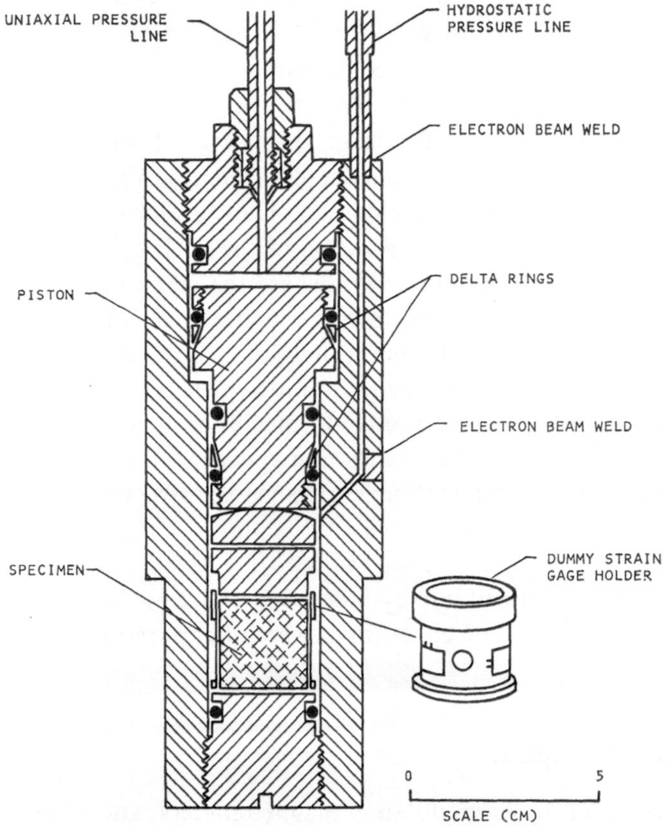

Fig. 2. High-pressure vessel.

Dummy strain gages are also inserted into the high-pressure environment to compensate for temperature effects and to provide flux cancellation for magnetic moments arising from gage currents.

4. Data and Discussion

The relationship between stress, σ, and remanence, J, over a wide stress range is well represented by (STACEY and BANERJEE, 1974)

$$\frac{J(\sigma)}{J(0)}=\frac{1}{1\pm S_r\sigma}$$

or

$$S_r\simeq\frac{1}{J(\sigma)}\frac{\partial J}{\partial\sigma}$$

where S_r is termed sensitivity of remanence. S_r depends upon the anisotropic magnetostrictions (λ) and susceptibilities (K). We distinguish between the sensitivities parallel ($S^{//}$) and perpendicular ($S^{\perp}$) to the direction of applied stress. For low

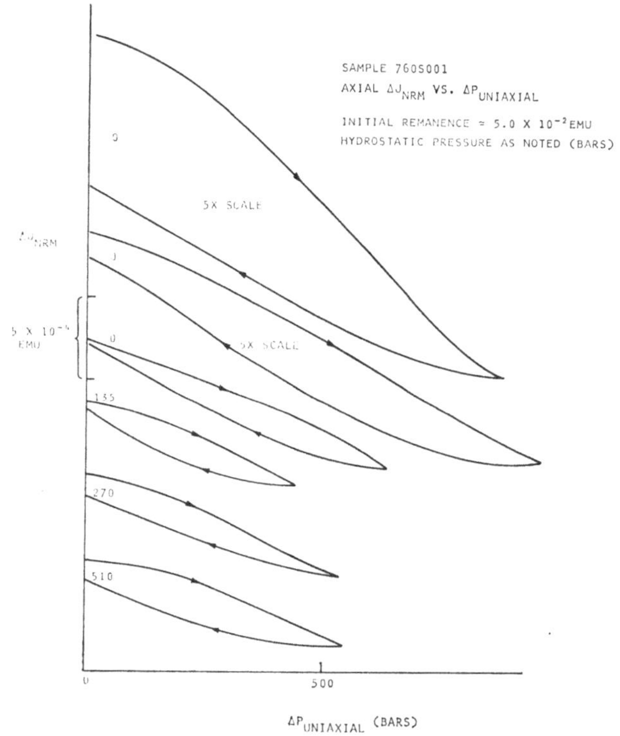

Fig. 3. Changes in the axial component (parallel to pressure axis) of NRM of sample 760S001 for some representative stress cycles. Top curve—1st stress cycle at zero confining pressure; 2nd curve—2nd stress cycle at zero confining pressure; 3rd curve—6th stress cycle; 4th, 5th and 6th curves—stress cycles at 135, 270 and 510 bars confining pressure, respectively.

stresses, simple theory (e.g., STACEY and JOHNSTON, 1972) predicts that:

$$S^{\perp} \simeq -\frac{1}{2} S^{//}$$

and that for titanomagnetites $J^{//}$ decreases with compression.

However, KEAN *et al.* (1976) have reported that $S^{\perp}/S^{//}$ can be significantly different from 1/2 in the case of multidomain grains. REVOL (1976) noted an increase in the component of remanence parallel to compression ($J^{//}$) during the initial part of a stress cycle on polycrystalline magnetite; similar observations on rocks have recently been reported by WYSS and MARTIN (1977). Stress sensitivities of NRM on the order of 10^{-4} cm²/kg have generally been found from laboratory experiments (e.g., KALASHNIKOV and KAPITSA, 1952; NAGATA, 1969).

Since our magnetometer measures along only two orthogonal axes (one parallel to the uniaxial compression axis and one perpendicular to it), we must, in general,

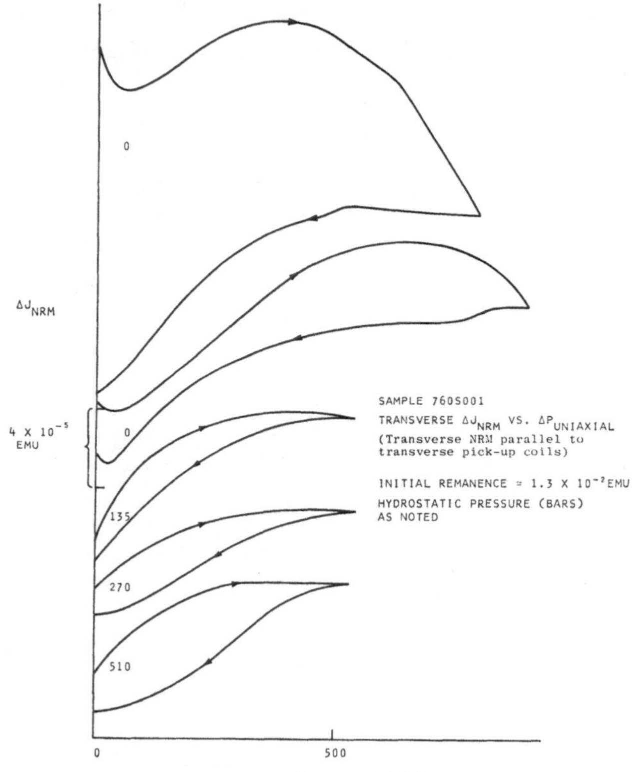

Fig. 4. Changes in NRM of sample 760S001 for one of two orthogonal components transverse to the pressure axis. For these measurements, the transverse component of the initial NRM was oriented parallel to the axis of the transverse pick-up coils. Top curve—1st stress cycle at zero confining pressure; 2nd curve—2nd stress cycle; 3rd, 4th and 5th curves—stress cycles at 135, 270 and 510 bars confining pressure, respectively.

perform two sets of pressure experiments to obtain a complete data set, particularly if rocks exhibit anisotropy. Although in many cases the rocks with which we are working appear texturally and mineralogically isotropic, anisotropy is introduced by the direction of magnetization, and may also result from preferred orientation of microcracks. The results of a series of runs on two specimens (760S001 and AN-2-275) are shown in Figs. 3 to 7. The experiments were carried out in field-free space (less than a few tens of gammas) on the NRM as it came from the field; no preliminary demagnetization was performed. Although strain gages were affixed for control, the volumetric strains are not shown since we are well within the linear elastic range.

Figures 3 and 4 illustrate the NRM stress sensitivities and demagnetizations for sample 760S001. The top three curves of Fig. 3 represent the change in axial NRM (parallel to compression axis) for the first three consecutive stress cycles at atmospheric pressures. Theoretically predictable behavior is observed; a decrease in NRM with compression occurs, and demagnetization decreases with each successive cycle. The bottom three curves represent consecutive cycles at increased hydrostatic pressure (i.e., 135, 270, and 510 bars). Interestingly, irreversibility is seen to increase

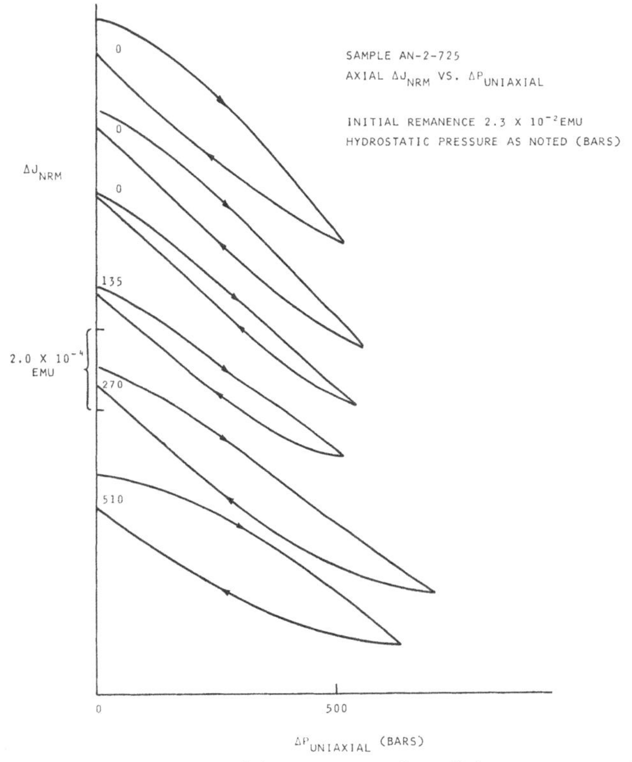

Fig. 5. Changes in the axial component (parallel to pressure axis) of NRM of sample AN-2-725 for some representative cases. Top three curves are the 1st, 2nd and 6th stress cycles at zero confining pressure; bottom three curves are the stress cycles at 135, 270 and 510 bars confining pressure.

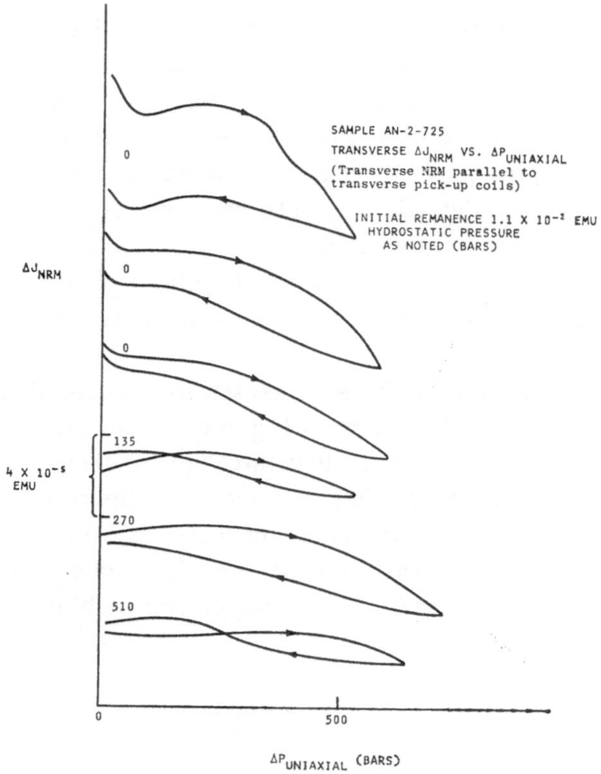

Fig. 6. Changes in NRM of sample AN-2-725 for one of two orthogonal components transverse to the pressure axis. The transverse component of the initial NRM was oriented parallel to the axis of the transverse pick-up coils. Top three curves are the 1st, 2nd and 6th stress cycles at zero confining pressure; bottom three curves are the stress cycles at 135, 270, and 510 bars confining pressure.

with an increase in the hydrostatic pressure. Thus additional demagnetization can be effected for a given stress cycle if the hydrostatic pressure is increased. This may reflect improved communication of stress to magnetic grains across grain boundaries as porosity is reduced and microfractures are closed. The top two curves in Fig. 4 represent the change in transverse NRM (measured parallel to the NRM) for two consecutive stress cycles at atmospheric pressure. For this component, the behavior is more complex than the simple theory predicts, with stress sensitivity changing sign during the stress cycle. The behavior becomes more predictable with additional cycling. Thus the unpredictable behavior may be related to the presence of a soft NRM component coupled with grain boundary effects. It is tempting to speculate on the role of secondary magnetic carriers which may occur along grain boundaries or as filler in microcracks.

The axial and two transverse sensitivities of NRM for sample AN-2-275 are shown in Figs. 5 to 7. The axial change in NRM exhibits expected behavior; as with sample 760S001, demagnetization increases with hydrostatic pressure. The transverse change in remanence, parallel to the NRM (Fig. 6) of this sample de-

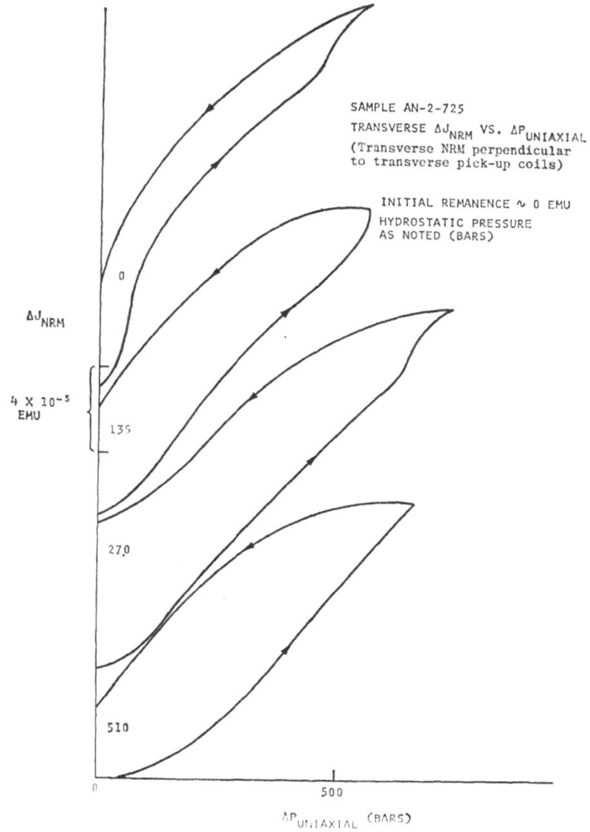

Fig. 7. Changes in NRM of sample AN-2-725 for the second of two orthogonal components transverse to the pressure axis. In this case the transverse component of the initial NRM was oriented perpendicular to the axis of the transverse pick-up coils. Top curve is the first stress cycle at zero hydrostatic pressure; the bottom three curves are the stress cycles at 135, 270 and 510 bars confining pressure.

creases, in contrast to the behavior in sample 760S001. Also some anomalous behavior appears in the initial stress cycles. Perhaps most interesting is the dissimilarity in behavior between the two transverse components (Figs. 6 and 7), in one case the component of NRM increasing and in the other case decreasing with applied stress. Additional experiments are necessary to test the validity of this result.

5. Conclusions

Cryogenic magnetometers are ideally suited for performing a wide variety of piezomagnetic experiments. Versatile high-pressure vessels can be constructed for insertion into the static sensing region of the magnetometer.

The measurement of stress sensitivity of NRM and the reversibility of initial remanence (demagnetization effect) are particularly important in the low stress range (0–500 bars) for pertinent rocks along the San Andreas fault.

Two quite dissimilar samples from widely spaced locations along the San Andreas fault have been studied. For both samples the NRM decreases monotonically parallel to the compression axis as simple theory predicts, with appreciable demagnetization observed during the first few cycles. This result can be compared to selected results of REVOL (1976) and WYSS and MARTIN (1977) which show an increase in the axial component during initial phases of a stress cycle.

Also apparent in our results is an increased irreversibility with increase in hydrostatic pressure (i.e., Figs. 3 and 4) which may be related to grain boundary interactions. The increase in hydrostatic pressure does not appear to appreciably affect the stress sensitivity.

The stress sensitivities perpendicular to the applied uniaxial stress show more complex behavior than the axial component. Initial stress cycles are complex with reversals in sign of the sensitivities. The transverse sensitivity, $S^{\perp}$, appears to be anisotropic, probably responding to the direction of NRM. The behavior perpendicular to the transverse component of NRM appears to correspond most closely to that predicted by the simple theory given in Section 4.

The erratic behavior observed during early stress cycles suggests that it is soft magnetic components which are responsible. It is not yet clear if these components occur under actual field conditions or have been subsequently induced. Also the presence of appreciable irreversible behavior must be cast in light of the assumption that these materials have been subjected to repeated stress cycling, in situ, during seismic strain build up and release along the San Andreas fault system. This suggests that either (a) stress cycling in the field has not reached the same peak levels as in the laboratory; (b) secondary NRM components have developed in near-surface rocks along the fault system which, due to time or present proximity to the earth's surface, have not yet been stress demagnetized; (c) 'laboratory' components of remanence have invaded the specimens; or (d) stress concentrations due to sample pressure vessel geometry may have induced 'non-representative' (of the low stress range) demagnetizations. Further experimentation will be needed to test the reproducibility of much of the piezomagnetic behavior reported here.

REFERENCES

BRUNE, J.N. and C.R. ALLEN, A low stress-drop, low magnitude earthquake with surface faulting: The Imperial, California earthquake of March 4, 1966, *Bull. Seismol. Soc. Am.*, **57**, 501–517, 1967.

BRUNE, J.N., T.L. HENYEY, and R.F. ROY, Heat flow, stress and rate of slip along the San Andreas fault, California, *J. Geophys. Res.*, **74**, 3821–3827, 1969.

CLARKE, J., Low-frequency applications of superconducting quantum interference devices, *Proc. IEEE*, **61**, No. 1, 8–18, 1973.

CLARKE, J. *et al.*, Tunnel junction DC SQUID: Fabrication, operation, and performance, *J.L. Temp. Phys.*, **25**, 99–144, 1976.

GOREE, W. and M. FULLER, Magnetometers using RF driven SQUIDS and their applications in rock magnetism and paleomagnetism, *Rev. Geophys. Space Phys.*, **14**, 591–608, 1976.

GRABOVSKII, M.A. and E.I. PARKHOMENKHO, On the change in magnetic properties of magnetite under the influence of high compressive stresses, *Izv. Akad. Nauk. USSR, Geophys. Ser.*, **5**, 405-517, 1953.

ISPIR, Y. and O. UYAR, An attempt in determining the seismomagnetic effect in northwest Turkey, *J. Geomag. Geoelectr.*, **23**, 295-305, 1971.

JOHNSTON, M. and F.D. STACEY, Transient magnetic anomalies accompanying volcanic eruptions in New Zealand, *Nature*, **224**, 1289-1290, 1969.

KALASHNIKOV, A.G. and S.P. KAPITSA, Magnetic susceptibility of rocks under elastic stresses, *Izv. Akad. Nauk. USSR*, **86**, 521-523, 1952.

KEAN, W.F., R. DAY, M. FULLER, and V.A. SCHMIDT, The effect of uniaxial compression on the initial susceptibility of rocks as a function of grain size and composition of their constituent titanomagnetites, *J. Geophys. Res.*, **81**, 861-872, 1976.

KERN, J.W., Effect of stress on the susceptibility and magnetization of a partially magnetized multidomain system, *J. Geophys. Res.*, **66**, 3807-3816, 1961.

LOUNASMAA, O.V., *Experimental Principles and Methods below 1K*, pp. 140-189, Academic Press, New York, 1974.

NAGATA, T., Magnetic susceptibility of compressed rocks, *J. Geomag. Geoelectr.*, **18**, 73-80, 1966a.

NAGATA, T., Main characteristics of piezo-magnetization and their qualitative interpretation, *J. Geomag. Geoelectr.*, **18**, 81-97, 1966b.

NAGATA, T., Tectonomagnetism, *Int. Assoc. Geomag. Aeron. Bull.*, **27**, 12-43, 1969.

NAGATA, T., Basic magnetic properties of rocks under the effects of mechanical stresses, *Tectonophysics*, **9**, 167-195, 1970a.

NAGATA, T., Anisotropic magnetic susceptibility of rocks under mechanical stresses, *Pure Appl. Geophys.*, **78**, 110-122, 1970b.

NAGATA, T. and H. KINOSHITA, Studies on piezo-magnetization: (I) Magnetization of titaniferous magnetite under uniaxial compression, *J. Geomag. Geoelectr.*, **17**, 121-135, 1965.

OHNAKA, M. and H. KINOSHITA, Effects of uniaxial compression on remanent magnetization, *J. Geomag. Geoelectr.*, **20**, 93-99, 1968.

REVOL, J., Laboratory magnetic observations related to earthquake prediction, M.S. thesis, 172 pp., University of California at Santa Barbara, 1976.

SMITH, B.E. and M.J.S. JOHNSTON, A tectonomagnetic effect observed before a 5.2 earthquake near Hollister, California, *J. Geophys. Res.*, **81**, 3556-3560, 1976.

STACEY, F.D., Theory of magnetic susceptibility of stressed rock, *Phil. Mag.*, **7**, 551-556, 1962.

STACEY, F.D., The seismomagnetic effect, *Pure Appl. Geophys.*, **58**, 5-22, 1964.

STACEY, F.D. and S.K. BANERJEE, *The Physical Principles of Rock Magnetism*, 195 pp., Elsevier, Amsterdam, 1974.

STACEY, F.D. and M. JOHNSTON, Theory of the piezo-magnetic effect in titanomagnetite-bearing rocks, *Pure Appl. Geophys.*, **97**, 146-155, 1972.

WYSS, M. and R.J. MARTIN, III, Tectonomagnetism and Magnetic Changes in Rock Prior to Failure, in *Proc. of Conf. II. Exp. Studies of Rock Friction with Application to Earth Pred.*, edited by J.F. Evernden, p. 449, U.S. Geol. Survey, Office of Earthquake Studies, Menlo Park, California, 1977.

YAMAZAKI, Y. and T. RIKITAKE, Local anomalous changes in the geomagnetic field at Matsushiro, *Bull. Earthq. Res. Inst.*, **48**, 637, 1970.

AEPS Vol. 1

Special Issue of Journal of Geomagnetism and Geoelectricity (Included in regular issues)

Proceedings of AGU 1976 Fall Annual Meeting, December 1976, San Francisco

ORIGIN OF THERMOREMANENT MAGNETIZATION

Edited by David J. DUNLOP

Contents TRM and Its Variation with Grain Size: A Review (R. DAY) / Single Domain Oxide Particles as a Source of Thermoremanent Magnetization (M.E. EVANS) / Domain Structure of Titanomagnetities and Its Variation with Temperature (H.C. SOFFEL) / The Demagnetization Field of Multidomain Grains (R.T. MERRILL) / The Hunting of the 'Psark' (D.J. DUNLOP) / On the Origin of Stable Remanence in Pseudo-Single Domain Grains (S.K. BANERJEE) / The Preparation, Characterization and Magnetic Properties of Synthetic Analogues of Some Carriers of the Palaeomagnetic Record (J.B. O'DONOVAN and W. O'REILLY) / Reduction fo Hematite to Magnetite under Natural and Laboratory Conditions (P.N. SHIVE and J.F. DIEHL) / Characteristics of First Order Shock Induced Magnetic Transitions in Iron and Discrimination from TRM (P. WASILEWSKI) / The Thermoremanence Hypothesis and the Origin of Magnetization in Iron Meteorites A.) BRECHER and L. ALBRIGHT) / Thermal Overprinting of Natural Remanent Magnetization and K/Ar Ages in Metamorphic Rocks (K.L. BUCHAN, G.W. BERGER, M.O. MCWILLIAMS, D. YORK, and D.J. DUNLOP) / Does TRM Occur in Oceanic Layer 2 Basalts? (J.M. HALL) / The Effects of Atleration on the Natural Remanent Magnetization of Three Ophiolite Complexes: Possible Implications for the Oceanic Crusl (S. LEVI and S.K. BANERJEE)

212 pp. (7 × 10) 1977 $ 24.50

AEPS Vol. 2

Supplement Issue of Journal of Physics of the Earth (Not included in regular issues)

Proceedings of the U.S.-Japan Seminar on Theoretical and Experimental Investigations of Earthquake Precursors

EARTHQUAKE PRECURSORS

Edited by C. KISSLINGER and Z. SUZUKI

Contents Earthquake Prediction-Related Research at the Seismological Laboratory, California Institute of Technology, 1974–1976 (J.H. WHITCOMB) / Research on Earthquake Prediction and Related Areas at Columbia University (L.R. SYKES) / Seismic Activities and Crustal Movements the Yamasaki Fault and Surrounding Regions in the Southwest Japan (K. OIKE) / The New Madrid Seismic Zone as a Laboratory for Earthquake Prediction Research (B.J. MITCHELL, W. STAUDER, and C.C. CHENG) / Anomalous Crustal Activity in the Izu Peninsula, Central Honshu (K. TSUMURA) / Recent Seismometrical Works in Japan (S. SUYEHIRO, M. ICHIKAWA, and K. TSUMURA) / Quiet and Violence in Horizontal Movement of the Crust (T. HARADA) / Anomalous Seismic Activity and Earthquake Prediction (H. SEKIYA) / Seismic Activity in the Northeastern Japan Arc (A. TAKAGI, A. HASEGAWA, and N. UMINO) / Observations of Changes in Seismic Wave Velocity in South Kanto District, South of Tokyo, by the Explosion-Seismic Method (T. KAKIMI and I. HASEGAWA) / Some Precursors Prior to Recent Great Earthquakes along the Nankai Trough (H. SATO) / Possibility of Temporal Variations in Earth Tidal Strain Amplitudes Associated with Major Earthquakes (T. MIKUMO, M. KATO, H. DOI, Y. WADA, T. TANAKA, R. SHICHI, and A. YAMAMOTO) / Gravity Changes Associated with Seismic Activities (Y. HAGIWARA) / Geomagnetism in Relation to Tectonic Activities of the Earth's Crust in Japan (N. SUMITOMO) / Precursory and Coseismic Changes in Ground Resistivity (T. RIKITAKE and Y. YAMAZAKI) / Geochemistry as a Tool for Earthquake Prediction (H. WAKITA) / Recent Laboratory Studies of Earthquake Mechanics and Prediction (W.F. BRACE) / Dilatancy of Rocks under General Triaxial Stress States with Special Reference to Earthquake Precursors (K. MOGI) / Possibility of a Great Earthquake in the Tokai District, Central Japan (T. UTSU) / Depth Constraints on Dilatancy Induced Velocity Anomalies (K.W. WINKER and A. NUR) / Seismological Precursors to a Magnitude 5 Earthquake in the Central Aleutian Islands (E.R. ENGDAHL and C. KISSLINGER) / Estimation of Future Destructive Earthquakes from Active Faults on Land in Japan (T. MATSUDA) / Some Problems in the Prediction of the Nemuro-oki Earthquake (K. ABE) / Responses to Earthquake Prediction in Kawasaki City, Japan in 1974 (H. OHTA and K. ABE) / Socioeconomic and Political Consequences of Earthquake Prediction (J.E. HAAS and D.S. MILETI)

304 pp. (7×10) 1978 $32.50

AEPS Vol. 3

Proceedings of the U.S.-Japan Seminar on Rare Gas Abundance and Isotopic Constraints on the Origin and Evolution of the Earth's Atmosphere

TERRESTRIAL RARE GASES

Edited by E.C. ALEXANDER, Jr. and M. OZIMA

Contents *EXPERIMENTAL STUDIES* A Mantle Helium Component in Circum-Pacific Volcanic Gases: Hakone, the Marianas, and Mt. Lassen (H. CRAIG, J.E. LUPTON, and Y. HORIBE) / Nitrogen to Argon Ratio in Volcanic Gases (S. MATSUO, M. SUZUKI, and Y. MIZUTANI) / Rare Gas Abundance Pattern of Fumarolic Gases in Japanese Volcanic Areas (O. MATSUBAYASHI, S. MATSUO, I. KANEOKA, and M. OZIMA) / A Review: Some Recent Advances in Isotope Geochemistry of Light Rare Gases (I.N. TOLSTIKHIN) / Abundances and Isotopic Compositions of Rare Gases in Granites and Thucholites (P.K. KURODA and R.D. SHERRILL) / Rare Gas Isotopic Compositions in Diamonds (N. TAKAOKA and M. OZIMA) / Rare Gases in Mantle-Derived Rocks and Minerals (I. KANEOKA, N. TAKAOKA, and K. AOKI) / A Comparison of Terrestrial and Meteoritic Noble Gases (O.K. MANUEL) / The Composition and History of the Martian Atmosphere (T. OWEN) *THEORETICAL STUDIES* Nuclear Components in the Atmosphere (T.J. BERNATOWICZ and F.A. PODOSEK) / Trapped Xenon and Cosmic-Ray Effects in Meteorites, in Lunar Sample, and in the Earth's Materials (K. SAKAMOTO) / Classification and Generation of Terrestrial Rare Gases (K. SAITO) / Earth-Atomosphere Evolution Model Based on Ar Isotopic Data (Y. HAMANO and M. OZIMA) / Terrestrial Potassium and Argon Abundances as Limits to Models of Atomospheric Evolution (D.E. FISHER) / On the Ambient Mantle $^{4}He/^{40}Ar$ Ratio and the Coherent Model of Degassing of the Earth (D.W. SCHWARTMAN) / Earth Degassing Models, and the Heterogenous vs. Homogeneous Mantle (R. HART and L. HOGAN) / Lead Isotope Constraints on the Early History of the Earth (R.D. RUSSELL) / Matter Accretion into the Solar System (S. HAYAKAWA)

230 pp. (7×10) 1978 $24.50

AEPS Vol. 4

Special Issue of Journal of Geomagnetism and Geoelectricity (*Included in regular issues*)

Proceedings of IAGA/IAMAP Joint Assembly, August 1977, Seattle, Washington

AURORAL PROCESSES

Edited by C.T. RUSSELL

Contents *TIMING OF SUBSTORM EVENTS* Pi 2 Micropulsations as Indicators of Substorm Onsets and Intensifications (G. ROSTOKER and J.V. OLSON) / The Use of Ground Magnetograms to Time the Onset of Magnetospheric Substorms (R.L. McPHERRON) / Substorm Onset in the Magnetotail (A. NISHIDA) *ELECTROMAGNETIC AND ELECTROSTATIC INSTABILITIES ON AURORAL FIELD LINES* A Review of Electrostatic Wave Measurements on Auroral Magnetic Field Lines (M.C. KELLEY) / Diffuse Auroral Precipitation (M. ASHOUR-ABDALLA and C.F. KENNEL) / Electromagnetic Plasma Wave Emissions from the Auroral Field Lines (D.A. GURNETT) / Theory of Electromagnetic Waves on Auroral Field Lines (J.E. MAGGS) *RAPID AURORAL FLUCTUATIONS AND ASSOCIATED PHENOMENON* Observations of Rapid Auroral Fluctuations (T. OGUTI) / Highlights in the Studies of the Relationship of Geomagnetic Field Changes to Auroral Luminosity (W.H. CAMPBELL) / Microburst Precipitation Phenomena (G.K. PARKS) *MECHANISMS FOR THE FORMATION OF AURORAL STRUCTURE* Observed Microstructure of Auroral Forms (T.N. DAVIS) / Birkeland Currents and Auroral Structure (H.R. ANDERSON) / Relationships between Particle Precipitation and Aurorol Forms (J.L. BURCH and J.D. WINNINGHAM) / Photometric Investigation of Precipitating Particle Dynamics (S.B. MENDE) / Generation Mechanisms for Magnetic-Field-Aligned Electric Fields in the Magnetosphere (C.-G. FÄLTHAMMAR) / Review of Auroral Currents and Auroral Arcs (G. ATKINSON) / Acceleration Mechanisms for Auroral Electrons (D.W. SWIFT) / Subject Index